William George Lock

Askja; Iceland's Largest Volcano

With a Description of the Great Lava Desert in the Interior, and...

William George Lock

Askja; Iceland's Largest Volcano
With a Description of the Great Lava Desert in the Interior, and...

ISBN/EAN: 9783744729819

Printed in Europe, USA, Canada, Australia, Japan

Cover: Foto ©berggeist007 / pixelio.de

More available books at **www.hansebooks.com**

ASKJA.

ICELAND'S LARGEST VOLCANO:

WITH A DESCRIPTION OF THE

GREAT LAVA DESERT IN THE INTERIOR:

AND A

CHAPTER ON THE GENESIS OF THE ISLAND.

BY

W. G. LOCK, F.R.G.S.,

TRANSLATOR OF "SPORTING LIFE ON THE NORWEGIAN FJELDS," AND
AUTHOR OF "JOTTINGS ON SPORT IN NORWAY."

Published by the Author at 16, Kingston Terrace,
Charlton, Kent.

1881.

J. S. LEVIN, STEAM PRINTER,
2, MARK LANE SQUARE, GREAT TOWER STREET,
LONDON, E.C.

CONTENTS.

CHAP.		PAGES
1.	PREFATORY	1—18
2.	TO THE LAVA FLOOD IN THE MÝVATNS ÖRÆFI	19—34
3.	OVER THE ÓDÁÐAHRAUN TO ASKJA	35—70
4.	THE PROBABLE GENESIS OF ASKJA, AND OF ICELAND	71—106

"Come, wander with me," she said.
"Into regions yet untrod:
And read what is still unread
In the manuscripts of God."

And he wandered away and away
With Nature, the dear old nurse,
Who sang to him night and day
The rhymes of the universe.'

(LONGFELLOW.)

ASKJA, ICELAND'S LARGEST VOLCANO.

CHAPTER I.

PREFATORY.

> 'Now the firm earth shakes like a frighted beast,
> And the island quakes from west to east;
> And seas of fire come from the rent,
> As though in ire by Heaven sent!'
>
> (JOHN MILL.)

IT is generally believed that *Hekla* is the chief volcano in Iceland; but that this is most certainly not the case will be shown in the following pages. In the desert interior of the island—the very fact of its being a volcano unknown even to the Icelanders themselves prior to 1875—stands a volcanic mountain whose vast proportions dwarf *Hekla* into utter insignificance. Its crater is between seventeen and eighteen miles in circumference, and consequently has an area of not less than twenty-three square miles! There is ample proof in the condition of the volcano that it has erupted time after time since the settlement of Iceland, but owing to its distance from the inhabited coastal districts, no earlier eruption than that of 1875

is recorded. Eruptions that occurred there were placed to the credit of other volcanoes, or were said to have taken place amidst the icy wastes of the *Vatna Jökull*, a glacier-covered mountain region south of the volcano, having an area of 3,000 square miles.

In the spring of the year mentioned, it may perhaps be remembered, all Europe was astonished by hearing that a large quantity of volcanic-ash had been wafted across the North Sea from the direction of Iceland and scattered over the Scandinavian peninsula as far inland as the central districts of Sweden. No news had been received from the Icelanders since the preceding November, the Danish mail steamers at that time not running in winter, and in March volcanic-ash brought the first tidings that a terrible eruption had taken place in that lone volcanic isle on the verge of the Arctic Sea. Ash had been wafted from Iceland to Norway before, but never in such large quantities nor so far inland as upon this occasion; and naturally considerable anxiety was felt in Denmark, until letters were received from the island, as to the fate of the inhabitants, the most fearful thinking it possible that Iceland, like 'Mark's Reef' in Cooper's charming novel, had sunk again beneath the waves whence it had been upbuilded by prehistoric volcanic agency.

The volcano which erupted in 1875 is the mountain above alluded to, the largest of a group named the *Dyngjufjöll* (Bower-mountains), the bower being the before-mentioned huge crater, which bears the name of *Askja* (Basket). This volcano stands nearly equidistant from the north, south, and east coasts, being situated a few miles to the east of the centre of the *Ódáðahraun* (Misdeed-lava-desert), the large lava desert in the interior of the island. The eruption, or more correctly series of eruptions, was of a most extraordinary nature; a terrific explosion in the heart of the volcano on the 4th of January first notified to the Icelanders that a volcanic eruption of some magnitude was about to take place. Some idea of the violence of this explosion may be formed from its effects: in *Askja* an immense mass of the lava deposits, five miles in circumference and of unknown but immense thickness, was disrupted, and sank bodily to a depth of over 700 feet into an abyss that must have existed below it in the bowels of the mountain! while the earthquake caused by the concussion was one of the most alarming ever experienced, immense rifts, ten to twenty miles in length, being opened in the north-east part of the island. The greatest disturbance at a distance from the volcano took place in a desert region known as the *Mjvatns Öræfi* (Midge-lake Desert), and here from the largest of the rifts molten lava welled forth throughout its whole length nearly continuously for four months after the earthquake. The rift commences at a spot distant thirty miles from the subsidence in *Askja*, and extends in a north-north-easterly direction for over twenty miles. It is evident from the nature of the eruptions at the two places that this lava came through a subterraneous channel from *Askja*, where prodigious quantities of pumice and volcanic-ash were ejected, but, it is believed, no lava.

Beyond some brief newspaper paragraphs announcing

that volcanic-ash, presumably from Iceland, had fallen in Scandinavia, and two others, one of which said *Hekla* had again erupted, and the other that an eruption had taken place amid the icy wastes of the *Skaptár Jökull*, on the western side of the *Vatna Jökull*, nothing was heard in England of this eruption until Mr. Wight, of Edinburgh, on his return home published in the 'Glasgow Herald' an account of his visit to the lava-flood in the *Mývatns Öræfi*, and of his futile attempt to ascend the volcano. This enterprising old gentleman tried so hard to accomplish his object, that he deserved to be the first alien to set foot in the huge crater of this volcano, but it was not to be. He says: 'At about three-fifths of the ascent I had to cave in. Sixteen stones weight, and years a bit beyond the middle of man's span, were not equal to the task.'

Notwithstanding that the well-known traveller, Captain Burton, with a work on Iceland actually in the press, sojourned for some time in the summer of 1875 within a two days' ride of *Askja*, and four hours' of the seat of volcanic disturbance in the *Mývatns Öræfi* (as also did the author of 'The Home of the Eddas,' an exceedingly pretentious work on the island, subsequently published), nothing more, in all probability, would have been heard in England of this volcano had not our adventurous fellow-countryman, Mr. William Lord Watts, chanced this summer to be successful in his second attempt to cross the *Vatna Jökull*—then, it is believed, crossed for the first time. From its icy wastes he saw smoke ascending from a large mountain amidst the lava desert to the northward; and notwithstanding he had encountered a snow-storm of several days' duration while crossing the *Vatna*, he ventured into the then virtually unknown wilderness of the *Ódáðahraun*, and forced his way to the volcano while considerable disturbance was taking place. In the November of the following year, he read a short paper before the Royal Geographical Society, and also published a very readable little work, entitled 'Across the Vatna Jökull.' Mr. Watts, in both paper and book, gave a very interesting and detailed account of his journey across the *Vatna* and the desert interior of the island; and also in his book a graphic account of what was going on both in *Askja* and the *Mývatns Öræfi* at the time of his visit to those places. His description of the volcano itself, however, is somewhat meagre, and in some very important particulars incorrect; but this is not to be wondered at, as at this time it was impossible to examine the volcano closely: 'The sides of the crater were evidently falling in,' says Mr. Watts, 'and huge wide cracks even where we stood, showed us that our position was not altogether a safe one; .... sometimes scarcely a minute elapsed between the roar of the stony avalanches.'

Having stated that Mr. Watts' account of the *Dyngjufjöll* and *Askja* is in some very important particulars incorrect, it will perhaps be as well, as their description in his paper read before the Royal Geographical Society does not quite occupy the whole of a page of 'The Proceedings of the Society,' to quote it, that those of my readers who are not in possession of the 'Proceedings' for 1876, may see what that gentleman had to say about the volcano, and be in a position to judge when they have perused this little work whether I am justified in my comment.

'I "cachéd" two days' provisions and proceeded to the Dyngjufjöll. I found these mountains to consist of a series of semi-detached sections, some of which had broken out in ancient times, and by their insignificant (!) lava-streams had helped to swell the widely-extending lava desert of the Odádahraun.

'These sections of mountains described a heart-shaped form upon the south, enclosing the Askja. This is a three-cornered piece of elevated land 4,000 feet high, about six miles long and three or four broad; it is easily reached by a glen upon the north-east side of the Dyngjufjöll. The principal crater, which erupted this year, is situated in the south corner of the Askja.

'The crater is enclosed upon the eastern and western sides by mountains rising in some instances 1,000 feet above the Askja plain; they appear shorn of their inner faces by the violence of the eruption, forming perpendicular cliffs of great height.'

I cannot pass this statement without observing that this is the case only on the eastern side of what Mr. Watts terms the 'crater' (the abyss formed by the subsidence, now the bed of a thermal lake), as the encircling mountain-wall, save to the east and south, is four and a half miles distant, the plain of *Askja* lying between. (See Map.) To continue:

'These cliffs are rapidly falling in. avalanches of stone occurring at frequent intervals, and had formed in two places steep slopes of pumice and débris, which it is possible to descend; all access to the floor of the crater is prevented, however, by an interior rim of the precipice immediately at the base of the heights.'

I here omit a paragraph eulogising the view over *Askja*, as it is not descriptive.

'This volcano, which, perhaps, we may be allowed to call the Oskjagjà (the chasm of the oval casket),* does not appear to have produced anything but pumice, mud, and water, copious floods of the latter having evidently flowed from its crater. It is curious to remark that although this volcano has ejected water, it is neither a glacial nor a snow-capped mountain, and it is situated more than 100 miles from the sea.'

I must again pause for a moment to observe that even a glance at the map accompanying Mr. Watts' paper would have prevented such a misstatement as that *Askja* 'is situated more than 100 miles from the sea,' it being, as before observed, and as will be seen by the map of Iceland herewith, almost equidistant from the north, south, and east coasts, the distance in an air-line being about sixty miles. It is also worthy of notice that Mr. Watts speaks of the mountain here in the singular, as being 'neither a glacial nor a snow-capped mountain,' not as 'sections of mountains,' and he is quite right in doing so, though hardly consistent; but he is quite wrong, as we shall presently see, in asserting that the mountain is neither glacial nor snow-capped; no doubt he meant to show that the floor of *Askja* was bare of snow and ice. He concludes his brief description as follows:

* I regret to have to add that the poetical name 'chasm of the oval casket,' which Mr. Watts here asks to be allowed to call the volcano, is unknown outside of his paper and book, Öskjagjá (Öskju—genitive of *Askja*, gjá—a rift) signifying Basket-rift. It will be noticed, moreover, that the word is spelt incorrectly by Mr. Watts.

'Leaving the volcano of Askja behind us and proceeding in a westerly direction, we perceived that the lava from the Odáda-hraun had entered the Askja upon its most western side, having run for a considerable distance up hill. Upon descending the Dyngjufjöll to the west, a broad plain, barren and black with sand and lava, opened before us; this was the Odáda-hraun.'

I should fancy that the Fellows of the Royal Geographical Society were about as wise subsequent as they were prior to the reading and publication of the above, as to the exact situation, magnitude, and appearance of *Askja*. It most certainly does not convey an impression to one's mind of a vast, almost circular crater over seventeen miles in circumference, lying at a depth of 800 feet at the least, within a mountain, its mountainous periphery broken by gaps to the present level of the floor of the crater in two places only; of a mountain that had been built up by the deposit of innumerable lava-flows around and above a volcanic-vent to the height of 2,300 feet above the plain of the *Ódáðahraun*, equal to 3,800 feet above sea level, and subsequently heightened by the upheaval of the immense masses of these lava-deposits, and of the substrata that now form *Askja*'s periphery, the *outer* circumference of which, at an altitude of 3,500 feet, cannot be less than twenty-four miles. The altitude given by Mr. Watts of the lava-covered floor of *Askja*, 4,000 feet, is correct within a hundred feet or so, as also is the height of the encircling mountain wall, 'rising in some instances 1,000 feet above the *Askja* plain;' in fact it varies in height from 800 to 1,500. I have never visited the western gap, through which Mr. Watts says the lava has entered from the *Ódáðahraun*, therefore I cannot say whether he is correct or not, but I have examined the eastern one, and there, lava that has issued in *Askja* has streamed out, not entered in; and, as the superficial lava in that crater lies at the least 2,300 feet above the level of the *Ódáðahraun*, it is quite certain, if Mr. Watts is right, that the lava which entered the western gap had ' run for *a considerable distance* up hill.'

Beside the mountain in which *Askja* is situated, the *Dyngjufjöll* consist of a number of minor semi-detached peaks, extending to the northward from the eastern part of the northern front of the chief of these mountains, and another large mountain on the N.W. (hitherto unshown on any map) whose eastern face is, I should judge, about five miles in length from S.S.E. to N.N.W., quite detached from the others, an isolated peak standing between it and the chief mountain.

With all its faults, Mr. Watts' description of *Askja* was the only one presented to the English reader prior to August of the present year, when a paper descriptive of the volcano by the author was published by the Royal Geographical Society; for, from the day Mr. Watts stood in its weird amphitheatre, until the author pushed across the deserts to it in the summer of 1878, it had remained unvisited, save by the Professor of Mineralogy at the Copenhagen University, Fr. Johnstrup. This scientist was sent by the Danish government in the summer of 1876 to ascertain the nature of the eruptions in the previous year. Upon his return, the

Professor read a paper before the Danish Geographical Society descriptive of the volcano and its doings; but of late years so little interest has been taken in the Icelandic volcanoes by English scientists, that not even a *résumé* of Professor Johnstrup's interesting and instructive paper ever appeared in England, much less a full translation. French scientists, however, had not so readily forgotten the volcano from which the previous year ashes had been scattered over the north of Europe, and a *résumé* of the Professor's paper appeared in a French scientific journal. Thus French geologists were enlightened, while their English contemporaries remained virtually in the dark as to the exact position and magnitude of a volcano which, there is reason to believe, marks the focus of volcanic activity in Iceland, and which has certainly played no unimportant part in the construction of the island.

I would here beg leave to digress for a moment to make a few comments upon the works recently published on Iceland, and to observe what a contrast to the pluck and energy displayed by Mr. Watts, almost out of provisions after a hazardous journey, pushing on footsore and weary over a pathless fire-blasted wilderness to an unknown active volcano, was the utter want of these characteristics shown by Captain Burton and Mr. C. G. W. Lock, the author of 'The Home of the Eddas,' both of whom, incredible as it may seem, actually sojourned, as previously stated, within a two days' ride of *Askja*, and four hours' of the seat of volcanic disturbance in the *Mývatns Öræfi*, without—as far as appears in their works—even visiting the latter place during the eruption. Mr. C. G. W. Lock and several others (I do not know whether Burton was of the number or not) certainly one day did set out for the *Öræfi*, but this is how the excursion ended. Upon the way, it seems, they met Mr. Watts' guide Páll, from whom, Mr. C. G. W. Lock says (p. 11), 'we learnt that the nearest volcanic vent was still two hours' ride further into the desert, and that the eruption of molten matter had ceased, to be followed with clouds of smoke. . . . The majority voted against continuing the ride forward, and we accordingly turned our ponies' heads.' What a vacillating set! I would have gone on alone had I been of the party; and therefore it is pleasing to be able to disclaim relationship with one so wanting in firmness of purpose as the author of 'The Home of the Eddas' appears to be from this and other incidents; notably the abandoning of the ascent of *Hekla* after journeying to its very base. The day turned out stormy, so instead of waiting until the morrow he turned his back upon this famous mountain, and in his book the whole matter is abruptly dismissed in the following words, which savour very strongly of the feeling that actuated the fox to make his memorable comment upon the quality of certain grapes hanging far out of his reach (p. 63): 'We scarcely felt disposed to go to the top simply for the sake of saying we had been there, and every other inducement had disappeared'!—the 1845 crater and other 'inducements,' it is to be presumed, had taken wing. There was no hurry, for he tells us at the conclusion of the chapter: 'There followed a week of enforced idleness' at *Reykjavik*, and yet our author 'had no intention of spending an hour in waiting for an eruption' of the *Geysir* upon the return journey from *Hekla*, so that he failed to see a sight for which most tourists cheerfully wait three or four days.

We could not expect an individual who thus naïvely describes his failures to carry out his undertakings, as if he thought it nothing remarkable to fail in whatever he attempts, and vacillancy a commendable characteristic, to accomplish much; and therefore it is not to be wondered at that he did not distinguish himself by exploring 'the unknown region to the south,' or cross the *Ódáðahraun* to *Askja* during his lengthy sojourn in Iceland. But we might have expected that he would have taken the trouble to ascertain, as his knowledge of geography appears to be so limited that he does not know the relative position on the Globe of the island about which he writes and the towns of Norway, and thus have saved the *exposé* of his ignorance in the following absurd statement based upon the fact that two newspapers are printed in *Akureyri*:—'Thus the Home of the Eddas may claim the honour of possessing the most northerly printing press in the whole wide world'!! One thing is certain, 'The Home of the Eddas' cannot claim the honour of possessing the most learned author in the whole wide world, for Tromsö, N. Lat. 69° 38′ (*Akureyri*, 65° 40′), I know from personal knowledge, had as far back as 1874 at least one newspaper printed there; and, as it is a town which even at that time had a population of 5,000, a High School (termed *Latinskole* as in *Reykjavik*), no less than three Banks, an excellent Museum, and other public buildings, it is surely of enough importance to warrant one in the belief that no one really an authority upon affairs Scandinavian could possibly be ignorant of its existence. However, I believe, Hammerfest, N. Lat. 70° 40′, claims 'the honour of possessing the most northerly printing press in the whole wide world.' I hope the information may prove useful to Mr. C. G. W. Lock, should his work reach a second edition; as also may a hint that it will be well in the future to abstain from penning such declamatory sentences upon other writers as the following, or he may again unwittingly plait a whip-lash for his own back: 'Ye gods, what a geographer and historian!'— 'Instances like these make one lean towards the practice . . . of appending writers' signatures . . . that outsiders may judge of his capacity for the office, and *know what value to attach to subsequent articles from one who has already shown signs of weakness.*'

To return to Burton, and his work. To judge from the gallant Captain's two bulky volumes he had drifted to the *ultima Thule* of literature, compilation; the bourne, alas! too frequently, of those afflicted with *cacoëthes scribendi*, when, as in this instance, they lack the energy to push into virtually unexplored regions, even when standing upon the very threshold. Whether Sir H. C. Rawlinson intended to be ironical or not, when at a meeting of the Royal Geographical Society he made the following comment upon Captain Burton's book on Iceland, he was exquisitely so: 'the well-known encyclopædic tastes and acquirements of the author enabling him in addition to the narrative of his journey to bring together a mass of information'! Truly, choice of words is everything! It being proverbial that encyclopædias are invariably a century behind the age, it is not to be wondered at that amongst the 'mass of information' brought together by Burton, there was only the following about a volcano that was active while his work was in the press, and the 'ubiquitous' Captain within a two days' ride of it: 'I wonder also how that queer elongated horse-shoe farther south, the "Askja" or "Dýngjufjöll hin Syðri" came to be laid out; but my knowledge of the ground does not allow me to correct the shape,' and, in the form of a footnote, a letter which went the round of the papers, stating that the eruption in January had taken place in the *Skaptár Jökull*. After one failure to 'plant a lance in Iceland' by ascending the mountain named *Herðubreið*, the 'veteran traveller' appears to have rested content with the laurels won in sunnier lands, for he did not again venture into 'the unknown region to the south,' as he himself terms the interior south of *Mývatn*.

It may be observed of Iceland that it is a country much written about but very little described; as notwithstanding there have been to my knowledge six books, besides the two criticised, published since the 1875 eruptions and a number of magazine articles, hardly a word is penned by any one save Wight and Watts, respecting the volcanic vents active in the winter and spring of that year. Although this appears strange at first sight, it ceases to be so when one bears in mind that the majority of the writers made but a two days' excursion inland to the *Geysir*, or confined themselves to other well-beaten and often before described tracks, diving—pretty deeply, too, some of them, to obtain matter for a volume or an article—into the pages of Hooker, Henderson, and other antiquated authorities for information respecting the eruptions and volcanic vents of Iceland, totally oblivious of, or ignoring the fact that volcanic eruptions take place in the island once or oftener in every decade (since 1821 thirteen eruptions are recorded, besides a number of earthquakes), not infrequently at spots where an eruption has never taken place before, or at least in historical times; *f.e.* the *Mývatns Öræfi* in 1875. One author brought out a work of close upon 300 pages, after having been ten days on the island, during which lengthy sojourn he safely performed the perilous journey inland as far as the *Geysir*, distant two days' easy ride from the capital!

In addition to the above comments upon the recent works on Iceland, I feel it incumbent upon me to repeat, that I think English scientists of late years have greatly neglected the volcanic vents of this island, and to observe that it is not over creditable to the Fellows of our Geological Society that not one has followed in the footsteps of Mr. Watts, after the Royal Geographical Society making known through its journal the existence of a vast active volcano in the interior of Iceland, which promised, if examined by competent geologists, to throw considerable light not only upon the volcanic system of that island, but also of Europe; it being well known that most of the great European earthquakes and volcanic disturbances have been either followed or preceded by terrific eruptions in Iceland. As it is, English scientists have rested content with the description of the volcano furnished by Mr. Watts, and, as a rule, utterly ignore its existence, and continue, year after year, to treat *Hekla* as the most important of the Icelandic volcanoes; whereas, to judge by the 'lay' of most of the lava-floods around *Hekla*, they do not appear to have flowed from that volcano, or even from rifts radiating from the volcanic vent of that mountain, but to have issued from rifts above a subterranean channel, which—if such exists and runs in the direction S.W. to N.E., indicated by the 'lay' of the lava—would intersect the *Askja* crater, and be in all probability connected therewith.

Upon my return in the autumn of 1880 after passing two summers in Iceland, during which I crossed the *Ódáðahraun* to *Askja* twice, I forwarded to the Royal Geographical Society the paper published in August last. It was impossible within the limits of a short paper to do justice to a volcano that has played a very conspicuous part in the formation of Iceland, and therefore I resolved to bring out the present work. I have spent considerable time and money, and spared no pains to make this monograph as complete as possible, venturing, as before said, across the great lava desert twice, besides devoting a considerable amount of time to the construction of a map of Iceland on a far larger scale than has hitherto been published in England, which shows every recorded site of volcanic disturbance and a number of places famous in Edda and in Saga.

I was greatly assisted in the construction of my map, and in my explorations upon my second visit to the volcano, by a small map of *Askja* and a tract of the *Mývatns Öræfi* by Lieut. (now Captain) Caroc, of the Danish Navy, who accompanied Professor Johnstrup to *Askja*; and as some slight return for the assistance, I will bring this somewhat discursive prefatory chapter to a conclusion by doing an act of justice to the Lieutenant, who is in danger of losing the credit of having made this survey. In 1844 a map of Iceland by an Icelandic cartographer, Björn Gunnlaugsson, was published in Copenhagen, and last year a reprint was issued by the Icelandic Literary Society, giving the *Dyngjufjöll*, and the other orological features of the *Ódáðahraun*, from the survey made by Lieut. Caroc in 1876, without the fact being stated, the map appearing wholly as the work of Gunnlaugsson; upon whose original map *Askja* is wrongly shown as an elongated oval space—the 'elongated horse-shoe' of Burton—environed within a narrow mountain wall having an opening to the N.N.E., and otherwise unbroken. When Gunnlaugsson in the summer of 1837 made an attempt to survey this portion of the interior, he was forced by fogs and snowstorms to abandon his project, and it was with the utmost difficulty his party succeeded in making their way back to a farm; and, although it is asserted that he made another and successful attempt the following year, I am inclined to doubt if he ever visited the mountains in the *Ódáðahraun*, for if he had I feel certain from the general accuracy of his map, that the orological features of this desert would have been more correctly delineated than they are. Therefore the honour of having been the first man to map out the *Askja* crater unquestionably belongs to Lieut. Caroc, who executed his task in spite of almost insurmountable difficulties; a snowstorm in Lat. 65° N., at an altitude of close upon 4,000 feet, would have daunted most men, especially as it lasted for thirty-six hours, but Caroc finished his survey of the crater in spite of it, though he was compelled to abandon the survey of the outer circumference of the mountain.

With reference to the *Dyngjufjöll* and *Askja*, as traced by Mr. Watts upon the large map of Iceland belonging to the Royal Geographical Society, they there appear as a modification of Gunnlaugsson's elongated oval, with the encircling mountain wall wrongly broken up into sections, there being, as will be seen by the map attached to this volume, but two gaps to the level of the lava-covered floor of *Askja*.

Upon the map attached to this work *Askja* is shown from Lieut. Caroc's survey, with some not unimportant additions by myself as to the features of the mountain in which this crater is embosomed.

Without further prefatory remark, I will now proceed to describe my journey in the summer of 1880 to the volcano, visiting *en route*, the rifts and lava-beds in the *Mývatns Oræfi*, and the nature of the eruptions at both places in 1875.

CHAPTER II.

TO THE LAVA FLOOD IN THE MÝVATN'S ÖRÆFI.

'Wide ruin spread the element around,
His havoc leagues on leagues may you descry;
And still the smouldering flame lurks underground,
And tosses boiling fountains to the sky!'

(UMBRA.)

IN the summer of 1880, after a pleasant but uneventful voyage in Messrs. Slimons' very comfortable steamer 'Camoens,' I landed at *Húsavík* (a trading port on the north coast of Iceland), in company with a fellow-passenger with whom I had become acquainted during the voyage out, the initial letter of whose name is H. Two days later, the 5th July, we set out for the interior in company with an Icelandic theological student, named Arni, as interpreter and guide—and a very decent fellow he proved. We were each mounted on a sturdy ambling pony, and in the lightest possible marching order, having no baggage whatever save our Macintoshes and a few rounds of ammunition, H. taking his gun with him to provide a duck for dinner upon a pinch, while I took a rifle in case we fell in with the herd of reindeer having its habitat in the *Mývatn's Öræfi*, and likewise my fishing-rod. With gun and rod anyone travelling in Iceland need

never dread a famine in the camp, feathered game and trout being abundant.

My companion not having been in Iceland before, we did not proceed direct to *Askja*, but went two days' journey out of our road to visit 'the lions of the north' on the way. These are (1) the Northern Geysir (Icl. *Uxahrer*, Oxspring, so named from an ox having once slipped therein), the largest of a group of hot-springs a few hours' ride from *Húsavík*, which 'spouts' a considerable quantity of water to a height of about twelve feet, at intervals of a few minutes; (2) *Ásbyrgi* (Gods'-rock), one of the most wonderful of Icelandic phenomena, an insulated triangular mass of rock, nearly a mile in width at the base, which gradually increases in height towards its apex, where its perpendicular cliffs are at least three hundred feet in height, its summit maintaining the downward slope towards the north of the surrounding country, from which it is cut off by a > shaped chasm, formed undoubtedly by the subsidence of the vast rocky mass that formerly lay between the cliffs of the Gods'-rock and those opposite; the faces of the cliffs are but little weathered, remaining as clean cut as if the subsidence had taken place less than a century ago, but nevertheless it is prehistoric; (3) the *Illjóðaklettar*, or Speaking Cliffs, insulated masses of rock and curiously formed craters in the wild valley of the *Jökulsá* (Glacier-river, so named as it rises amid the *Vatna Jökull*), which appear like vast ruinous castles, and so perfectly do they echo back sounds, one can fancy them the home of Mocking Genii; and last, though not least, the *Dettifoss*, the most famous of Iceland's falls—on the whole, the grandest tour that can be made in the island.

This *détour* entailed two nights' very rough accommodation, the first being passed not over comfortably at *Ás*, the farmhouse near the Gods'-rock, and the second still more uncomfortably at an abandoned and ruinous hovel near a lake in the midst of the *Öræfi* named *Eilifsvatn*; whence four hours' ride over the desert waste brought us, early on the morning of the third day, in sight of the 1875 lava-flood. Seen from a distance it presents the appearance of a number of vast black heaps, as if some Titanic foundry had here shot its refuse slag and clinkers. Riding in a south-easterly direction straight towards the lava we, when within a mile or so of it, found our further progress barred by one of the rifts opened by the earthquake on the memorable 4th January. This we had to follow in a southerly direction until we struck the excellent newly-made road, the only thoroughfare across country from east to west, the other routes being round the north and south coasts. By keeping to this we were able to pass the numerous deep narrow cracks and fissures running parallel with the rift whence the lava issued, this new road having been constructed subsequently to the eruption, the old one having been blocked by the flow of lava.

Our route over the pathless *Öræfi* had been a somewhat difficult and dangerous one, for several times we were compelled to dismount and lead our ponies over lava-flows lying in hollows where subsidences had taken place between old earthquake rifts. When I visited the scene of the eruption here in 1878 I learnt what dangerous ground it was to traverse, for my pony placed both his forefeet in a crack in the earth, which my guide's pony had passed in safety, the thin turf above being unbroken, and, as a fire-worshipper should, I lay prostrate upon my face for a short period upon entering the hallowed precincts of this recently-acquired domain of the Fire King. Luckily I was not hurt, neither was my steed, being more fortunate in this respect than the Governor of the island, who, when he visited the scene of the eruption, lost one of his spare ponies. The poor brute fell bodily into one of the rifts, and a man had to be lowered down to cut its throat, it having in its struggles soon jammed itself inextricably fast between the rocky walls of the rift at some depth. A fine buck reindeer also managed the autumn before last to get into a rift, and the peasants, when searching for their sheep, found the animal suspended by his wide branching antlers in a most emaciated condition. He was extricated and led for some distance in the direction of *Reykjahlið*, but when he recovered his strength somewhat, he gave his captors so much trouble, that they ultimately killed him.

Approaching the 1875 lava, we see that on the north it lies for some distance in a hollow, similar to the older ones we had passed on our way from *Eilífsvatn*, formed by a subsidence,* during the eruption, of a tract three to four miles in length between two deep

---

* It is said by the people living around *Mývatn*, who scour the *Ódáði* every autumn in search of strayed sheep, that a

parallel rifts, which run northward for several miles, and that the lava stretches away southward farther than the eye can follow it, when one stands upon the level ground. The depth of the subsidence is greatest near the lava, from which the surface of the sunken tract slopes gradually up to the level of the Öræfi, the rifts decreasing in size as the depth of the subsidence becomes less.

Leading our ponies, we followed the western rift southward between it and the lava. It is somewhat similar in appearance to the well-known Allmen's-rift and Raven-rift, near Þingvellir, but narrower and far deeper; its depths not having been filled in with débris and soil, as is the case with the two mentioned. A quantity of snow, remaining from the preceding five winters, lay at the bottom of the rift, no ray of sunlight ever penetrating there. It cannot longer be doubted that the Allmen's-rift and Raven-rift were formed by the breaking away and subsidence of the tract between them, in the same way as these two recently-formed rifts in the Öræfi.

The lava now forms a bed tending from S.S.W. to N.N.E., varying greatly in breadth, and said to be over twenty miles in length. To judge from its appearance, a vast quantity, most intensely heated,

---

similar subsidence is to be seen at the southern end of the lava-bed. I cannot say whether this is true or not, having only ridden southward along the lava for a distance of eight miles. I was told in 1878 that the rifts seen on each side of the northern end of the lava extended its whole length, but this is not the case.

must first have welled forth from a rift running down the centre of the bed for nearly its whole length. This, owing to the comparatively level nature of the ground, spread freely westward and eastward wherever slight depressions existed in the plain of the *Öræfi*. In places these arms cover tracts several square miles in extent. The largest is on the western side, about the centre of the bed, and covers an area of some six or seven square miles. This first eruption of molten rock congealed into a rugged sheet, twelve to fifty feet in thickness, very clinker-like along its borders, which are fragments of the earlier cooled fringe of the molten flood that were borne forward by that behind. It is evident from the veritable chaos of huge masses of lava piled up in places down the centre of the bed, above the rift whence the lava issued, that eruption after eruption took place, each lava-flow being congealed into a layer of rock, which in its turn was subsequently upheaved and shattered by a later outburst, in whose fiery embrace the huge jagged masses of the torn-up bed were borne along partly embedded, and from which, now that the later of those molten floods are also solid rock, they project at all angles. These masses are also built up into several groups of rude cone-shaped craters having an altitude of from one hundred to two hundred feet.

By the aid of my hands, protected by thick woollen mittens that they might not be cut by the lava, I made my way with some difficulty across to a group of craters thus built up, at a spot about a mile and a half from the northern end of the lava, to get an idea of the extent and general appearance of the bed, and ascertain whether the lava had issued, as I surmised, from a continuous rift, or from older volcanic cones, a number of which stud the *Öræfi*. I had surmised rightly, for nowhere could I see evidence of any molten matter having streamed down the outer walls of the craters, or that the craters themselves were older than the lava. They appeared to have been merely vents maintained through the lava-flows at intervals above the rift for the escape of heated vapours from below, in which, at times, molten matter had risen a few feet ere it found an outlet at a lower level; which the later lava-flows appear to have done by forcing their way under and through the congealed earlier ones in a most extraordinary manner, reducing the under surfaces of the older, wherever they had been in contact with the fiery floods, again to a molten state, and upheaving immense masses. Radiating from the craters, I also noticed numerous deep fissures formed by the contraction of the lava while cooling, and deep abysses, plutonian in their gloom, where huge slabs had fallen in after the molten lava had flowed away from beneath. No heated air was ascending, so that in five years the whole mass appeared to have quite cooled

Owing to the erratic manner in which the lava has spread around and been piled up down the centre of the bed, it is not an easy matter to compute the cubic contents of the enormous molten mass that here issued in 1875; but I believe I shall under rather than over estimate if I take the length of the bed at twenty miles, and give it a mean width of five miles—

the vast sheets forming the arms being allowed for those parts of the main bed of far less width, and a thickness of one hundred feet—not too great a thickness when the vast quantity piled down the centre of the bed is taken into consideration, as well as the fact that all depressions in the area covered have been filled in; and thus computed it amounts in round numbers to over thirty-one thousands of millions of cubic feet. This lava-flood, therefore, is as nearly as possible twice as large as that from *Hekla* in 1845, which has been computed to contain 14,400 millions of cubic feet (Danish), and it dwarfs into insignificance the lava-streams ejected during the eruptions of Vesuvius in 1794 and 1855, which have been computed at 730 and 570 millions of cubic feet respectively. I believe that only once since the settlement of Iceland has a larger one flowed forth, and that was in 1783, from rifts which opened in, and north of the *Varmárdalr* (Warm-river-valley) at the foot of the *Skaptár Jökull*, on the west side of the *Vatna*. Burton and other writers state that this lava flowed from the *Skaptár Jökull*; but an Icelander, Herra Thoroddsen, who contributes a paper on the Icelandic Volcanoes to the October number of 'The Geological Magazine' says: 'These eruptions are erroneously stated to have taken place in *Skaptárjökull*, where an eruption has never yet occurred.' But what is remarkable, Herra Thoroddsen, with strange inconsistency, in the same paper credits the *Jökull*, under its other name of *Síðu Jökull*, with an eruption in the year 1389, and one in its neighbourhood in 1753.

Herra Thoroddsen's paper bears further evidence of having been very carelessly compiled, amongst other things, Danish measure is given for English, *Askja* 'measuring one square geographical mile in extent,' and the *Ódáðahraun,* '60 square geographical miles in extent.'

Riding along the edge of the rift on the western side of the lava, the almost perpendicular and clean-cut face of its wall, from whence the tract broke away, affords an interesting glimpse at the geological history of this part of the elevated plateau in the interior of Iceland. The altitude here, by Aneroid, is 1,300 feet above sea level, and to the depth of the rift, quite 200 feet in places, we see that the plateau is built up of a succession of beds of basaltic lava, lying horizontally stratum upon stratum, with thin layers of clinker-like fragments marking the divisions between. The strata vary in thickness, but not, as far as I could see, in appearance, there being no apparent difference in the density and colour of the substrata at a depth of sixty to seventy feet—to which depth I descended into the rift—and the more superficial ones, the lava being very dense at the bottom of each stratum, becoming less so and cellular towards the surface. The lava is generally of that ashy-gray hue, peculiar to basaltic lava and basalt, but in places there are bright red and yellow patches, caused doubtlessly by the oxidation by intense heat of minerals in the lava. In places the superficial stratum is columned by vertical fractures, possibly the effects of frost; and on the whole these deposits present more the appearance of a tertiary formation of basalt, than floods of basaltic lava, which have issued subærially in comparatively recent times. These strata are covered with earthy deposits to a depth of from three to four feet only, a rusty brown loam about a foot in depth rests upon the lava, then a thin layer, about three inches in thickness, of white pumiceous earth is met with, and above this is more of the rusty brown loam, which with black volcanic sand, forms a poor soil that nourishes the scanty vegetation that clothes the moorland. The layer of pumiceous earth, I would here observe, is found spread over the whole of the north of Iceland, even on mountain ridges 2,500 feet high it is to be seen wherever the earth covering it has been washed away or otherwise removed.

Whence came the floods of basaltic lava beneath the *Öræfi*, and at what period of the island's history? That they were ejected subærially is certain, and that the air above them as they spread one over the other must have been more intensely heated than that in the interior of a furnace is equally certain, otherwise they would not have spread around so evenly or so far. It is far easier to ask these questions than to answer them, but I believe that these vast sheets of lava welled forth from a huge volcanic vent underlying *Askja*, whose outlet has been narrowed down by the accumulation of these deposits to the present dimensions of that crater. The comparatively level plateau which now forms the greater part of the interior—the more superficial strata of which are exposed by earthquake rifts in every direction—is unquestionably a more recent formation than other parts of the island, that has been upbuilded in post-tertiary times by floods of igneous rock, chiefly from a central vent. As I purpose in the concluding chapter of this work to describe the formation of Iceland, I will not further digress here than to observe that the coastal region mainly consists of semi-detached flat-topped mountain masses varying little in altitude, about 2,000 feet, that appear to be fragments of a far older, more elevated, and extensive plateau than the one now forming the interior, though of very similar formation; fragments of one which, in all likelihood, was riven asunder during the disturbances of the glacial epoch, the greater portion then sinking beneath the sea. *Herðubreið* and several of the ice-clad *Jöklar* in the interior, as well as the mountain masses of the coastal region, are portions of this older plateau; and these, doubtless, at the beginning of the post-tertiary epoch formed a complex of islands very similar in appearance to the Faroes of to-day, but scattered over a greater area, and have been united and formed into one large island by subsequent outbursts of igneous rock from the volcanic vent in their midst.

I am able to give the following particulars of the eruption in the *Öræfi*, for which I am indebted to Jón, the intelligent son of the farmer at *Reykjahlíð*, the nearest inhabited house to the scene of the outbreak; where, by-the-bye, I stayed in August, 1880, when I bagged four reindeer, the first of these animals, it is believed, that have fallen to an Englishman's rifle in Iceland.

Lava was first seen issuing in the *Öræfi* on the 18th February, forty-five days after the earthquake, but it is probable the fiery flood commenced to stream forth immediately after, no one having crossed from *Reykjahlíð* to the eastward or *vice versâ* during that period; but little travelling naturally being done in Iceland during the winter. For nearly four months the lava continued to stream forth more or less freely, and then ceased to flow until the 15th August, when a smart shock of earthquake was felt, and a slight eruption of ashes and volcanic bombs took place from the rift at its northernmost end; the lava flow recommencing and continuing for several days. This eruption was witnessed by Mr. Watts, who in his work 'Across the Vatna Jökull,' very graphically describes the scene. The season of the year when the lava first burst forth, and the intermittent manner in which it issued, fully account for the chaotic confusion which the masses of the rapidly congealed earlier flows upheaved by the subsequent ones now present.

Having gratified our curiosity by exploring this, the largest lava flood that has issued in Iceland during the present century, we mounted our ponies and turned their heads in the direction of *Reykjahlíð*. We had not been in bed nor had a 'square' meal since leaving *Húsavík*, consequently we were anxious to get under the hospitable roof of the unjustly maligned Pétur Jónsson, where I had been made very comfortable in 1878.

West of the new lava, a capital newly made road

now crosses the barren moorlands and sandy wastes of the *Öræfi* for several miles, and over this we did not spare our ponies, a little over two hours' gallop bringing us to the lava-beds lying east of the *Námafjall* range of mountains bordering *Mývatn*. Here the made road came to an end, and we had to allow our ponies to slowly pick their way over the deeply fissured lava, the cracks and crannies in which are complete traps for tired out ponies, which do not pick their way so carefully as when fresh.

At the base of the *Námafjall* (Solfatara-mountain), we dismounted, and leaving our ponies in charge of Arni, H. and I started off to see the boiling mud wells in the solfatara at the base of the mountain. These boiling mud wells are the largest in Iceland, but as they have been more or less correctly described by at least half-a-dozen previous writers, I will only devote a few lines to their description. Between the *Námafjall* and the old lava beds to the eastward, resting upon a hot, white, viscid clay, is a plain of light coloured mineral earths, about half-a-mile in width from east to west by one and a-half in length from north to south. The earths, where wind and sun-dried, form a crust, in places capable of bearing a man, but there are large patches where it is unsafe to go, the crust being but an inch or so in thickness. The plain is studded with a number of low cone-shaped hillocks, from whose apices jets of steam ascend. These jets of steam mark where sulphur sublimation is taking place. The boiling mud wells lie on the eastern side of the plain, about half-a-mile from its northern end, in a bank of baked mud raised a few feet above the level of the plain, partly surrounded by a pool of hot water. Prodding the treacherous ground with our whip-handles, that we might not inadvertently step on to a soft spot and break through into the scalding hot clay, we crossed to the mud wells. These numbered at the time of our visit twenty-seven, and were in groups in three crater-like basins in the mud bank, a short distance from each other. The basin which held the principal group was about twenty feet deep and seventy in diameter, and contained at this time seven wells, the largest about thirty feet in diameter, and the smallest about the size of an ordinary pitch-kettle. The boiling mud was as black as ink and of the consistency of porridge, and might with propriety, I think, be termed 'Hell Broth.' It was in a constant state of ebullition, large quantities of that in the largest well being every few minutes ejected with a roar to heights varying from six to eight feet. Similar 'spouts' occurred from the largest well in each of the other groups. The wells in each basin are separated from one another by walls of dried mud: and I think it certain that the wells frequently vary considerably in size and number, the dividing mud walls within the basins constantly undergoing the process of destruction and re-construction, being at one time builded up by sediment ejected by the 'spouts,' and at other times boiled away by the liquid heated matter, as during our visit, one of the walls separating two of the smaller wells toppled over, and the two became one.

TO THE LAVA FLOOD IN THE MÝVATN'S ÖRÆFI. 33

Thus it is possible that one day there may not be more than one or two wells in a basin, while on another there may be a dozen, and if so, the difference in the number and dimensions of these 'Makkaluber' given by different writers is accounted for.

These boiling mud wells are a very interesting, though certainly not a beautiful phenomenon, and being situated at the junction of extensive lavabeds with a solfatara where hundreds of fumaroles are fizzing away, with a range of volcanic mountains, which also show signs of activity, within a few hundred yards to form a fitting background, they and their surroundings are very suggestive, as an American said of the Yellowstone region, 'of the Inferno very thinly crusted over.' 'Umbra's' lines heading this chapter, though not intended to do so, describe the scene to perfection.

Remounting our ponies a half-hour's ride through the pass in the mountain range, the *Námaskarð*, brought us in sight of the famous Midge Lake. This sheet of water and its surroundings seen from the pass on a clear day, similar to what we were favoured with, is magnificent. Rounding a projecting spur of the mountain on the north side of the pass, the circumscribed view of an unusually gloomy defile, wherein not even a blade of grass clothes the acclivities, changes as if by magic to one over a wide expanse of lake, environed on the south and east with volcanoes, and studded with volcanic isles, miniature quiescent Strombolis whose weather-worn cratercones rise from bases green with a prolific growth of

D

angelica and grasses; their verdure presenting a pleasing contrast to the bristling lava-floods which fringe the lake. I have ridden through this pass a goodly number of times, but whenever I do so the sudden change always reminds me of the transformation scene of a pantomime, whose brightness and glories are invariably preceded and enhanced by a scene gloomy in the extreme. *Mývatn* is the second sized lake in Iceland, *Þingvallavatn* being the largest, and is said to have an area of thirty square miles, and to lie 900 feet above sea-level.

Professor Johnstrup says of the *Mývatn* district: 'There is no region in Iceland which has such a thoroughly volcanic character as the district east of *Mývatn*, in this respect it deserves the appellation classic; it is The Fire Focus of the North where hundreds of volcanoes stand silent witnesses of catastrophes respecting which history hath penned not a word.' It richly deserves the title given, 'Fire Focus of the North,' but I believe the true focus of volcanic activity in Iceland is *Askja*, and that the volcanic vents east of *Mývatn* are upon a channel running therefrom northward.

From the pass less than an hour's ride, first through a solfatara very similar to the one east of the range, and subsequently over an extensive bed of lava full of rifts and chasms, brought us to the door of *Reykjahlið*, and in a very short time we were seated in the clean and very comfortable 'guest-room' before a dish of delicious boiled char, which during the preceding night had been disporting within the lava grottoes laved by the waters of Midge Lake.

CHAPTER III.

OVER THE ÓDÁÐAHRAUN TO ASKJA.

' But he whose weary step hath traced
Iceland's mysterious awful waste,
Whose eye the Ódáðahraun hath viewed
Can tell thee what is solitude!

Mindful of the past, the timid swain
Nurtured near the dread domain,
Hath peopled the caverns of the Hraun
With lawless sons from outlaws born!'

(Adapted from Mrs. Hemans.)

ARLY the next morning we set out for *Svartákot* (Black-river-cot), a farm situated on the border of the *Ódáðahraun*, whose owner acted as my guide to *Askja* in 1878. The cot stands close by the eastern shore of the *Svartávand* (Black-river-lake) on the northern bank of the river issuing therefrom. Both river and lake are well stocked with char and trout.

At Arni's request we went a little out of our way to his home *Skútustaðir* (Skútu's-stead), a parsonage attached to a church at the south end of *Mývatn*. Here, as no less than five couples were to be married, we somewhat reluctantly consented to stay several hours that Arni, who was more or less closely related to at least half the parties, might witness the ceremony. Space cannot be spared to describe the weddings, or the proceedings after, much as I should like to do so, as I must 'speed the pen,' otherwise, reader, you and I will never, figuratively, reach the volcano amidst the Misdeed-lava-desert.

It was six in the evening before we got away, so to make up for lost time, Arni essayed a 'short cut' which resulted in a flounder through a bog for nearly three hours, and, as we contrived afterwards to lose our way on the moorlands west of the mountains south of *Mývatn*, it was five o'clock the next morning before we reached our destination, which is distant but six hours' ride from *Skútustaðir* by the regular horse track.

We went to bed immediately upon our arrival, and slept until after midday. Four hours later saw us again in the saddle and, with Einar the farmer as guide, on our way to *Askja*. We took nothing with us but food for two days, and a spare pony laden with two large knitted woollen sacks, crammed full of hay as fodder for our steeds.

An hour's ride across level moorland, alive with willow grouse (*Lagopus subalpina*. Nils.), brought us to a considerable river named the *Suðrá*—South river (?)—a tributary to the Shivering-flood. We had followed a dry watercourse, formerly the channel of a small stream flowing from a tarn. I was told by Einar that for some years from fifteen to twenty of his sheep were drowned annually in this stream, the unfortunate brutes, it appears, when they came to drink, frequently slipped in, and were unable to get out, the sandy banks being concave, where they had been hollowed out by the running water, with the matted roots of the grass and bushes growing above overhanging considerably. One year—1872 I think—Einar lost seventy! This, he at last considered, was too high a tax to pay; so he finally did what ought to have been done years before, as it was only a matter of a few hours' work, dammed the stream at its outlet from the tarn. What was somewhat extraordinary, the water in the tarn after rising considerably above its former level, suddenly fell to this again, its overflow having found a subterranean outlet. I merely mention this incident to show what a long-suffering, patient, and apathetic being, an Icelandic farmer is; and yet Einar is an exceptionally enterprising one.

As the *Súðrá* flowed from the south-east, the direction in which we wished to proceed, we followed its course for about four miles, passing through an extensive sandy waste—not lava-desert as shown on Gunnlaugsson's map—bordering the *Ódáðahraun* on the north-west, named after the river, *the Súðrásandr*. The river, I discovered, was fed with the surface water from the *Ódáðahraun* in a somewhat remarkable manner, beautifully clear springs bubbling briskly forth in innumerable shallow pools in the sand. The phenomenon is easily accounted for. The surface lava in the desert lying at a higher level than the *Súðrásandr*, the melted snows and rains percolating through for centuries have worn channels in the porous lava and through these the surface water now flows until it reaches the sand, where it rises as we have seen.

We quitted the river at a still more remarkable pool, one in which the order of things was reversed, the water being in rotary motion around a small hole at the bottom, a few inches in diameter, through which it was descending to some underground channel. Here the ponies were allowed to drink to their hearts' content, their riders likewise imbibing as much as possible, as in a few minutes we should find ourselves amidst the lava where no water would be met with for at least six hours.

A brief description of the *Ódáðahraun*, which we were now about to cross, will doubtless prove interesting. Its area, as before stated, exceeds 1,100 square miles; and it forms part of a desert region lying between the *Jökulsá*, draining the *Vatna*, on the east and the *Skjálfandafljót* on the west, stretching south from the range of mountains encircling *Mývatn* on the south and east right to the glaciers of the *Vatna Jökull*, which extend to the coast; a black, fire-blasted, scarcely passable wilderness of lava and volcanic sand, at least two thousand square miles in extent. The surface lava of the *Ódáðahraun* has an altitude of about 1,500 feet, and it consists chiefly of countless lava-floods, varying greatly in age, some being thousands of years old and clothed with lichen, while others are as black and new-looking as those which flowed from the mountains east of *Mývatn* a century and a-half ago. The newer lava-floods in the vicinity of the *Dyngjufjöll* on the north have flowed from rifts that have been time after time torn in *Askja's* encircling mountain-wall, and at its base; while the south-western portion of the desert, according to Mr. Watts, is covered with those which have flowed from the *Trölladýngja* (Trolls-bower), a volcano lying about fifteen miles distant from *Askja* in that direction. Mr. Watts is the only man, I believe, save his guides, that has visited this mountain of late years, or crossed the desert between it and the *Dýngjufjöll*, and he says, p. 98 of 'Across the Vatna Jökull,' at the summit he 'found a small but perfectly formed crater, about 500 yards in circumference, but of no great depth, while in the centre rose a ridge of burnt lava, which gave the mountain the black tufted appearance I had noticed in the distance.' This crater-'bower' therefore is an utterly insignificant one compared with the 'bower' within the monarch of the *Dýngjufjöll*. Mr. Watts furthermore says the *Trölladýngja* is 'nothing but a huge mound of basaltic lava, partially covered with snow, rising by a very gradual slope to about 4,000 feet above sea level.' I may as well here observe that notwithstanding this mountain is named on Gunnlaugsson's map *Skjaldbreið eða* (or) *Trölladýngja*, and Mr. Watts and other writers invariably designate it by the first name, the Icelanders, or at least the people living around *Mývatn*, do not know it by that appellation, but always call it *Trölladýngja;* therefore I have done likewise, it being less likely to cause confusion, there being another *Skjaldbreið*, near *Þingvellir*.

A large extent of the desert in the north is covered with a prodigious lava-flood from a crater near the *Fremrinámar* (Farther solfatara), while in the north-

west it chiefly consists of outbursts of lava from huge
rifts similar to the one that issued in the *Mývatn's
Öræfi* in 1875, and a smaller one a few miles northeast of *Hekla* in 1878. The lava which has welled up
in this manner is the newest looking and most rugged.
The oldest lava that I saw appeared to be the surface
of the last of the veritable oceans of molten rock
which at one time overspread this portion of the interior of Iceland. We saw some of them, it will be
remembered, bared by the rifts in the *Mývatn's Öræfi*.
Occasionally tracts of this are met with between the
newer lava-floods, where the lava has congealed so
evenly that its surface is as level as a paved footpath;
and such places being swept clear of sand by the
winds, the traveller's pony, which usually in this
desolate region proceeds very spiritlessly, can be
forced into a trot. I noticed also tracts where this old
lava had formed innumerable flattish dome-shaped
bubbles starred with deep fissures, caused by the contraction of the mass when cooling, and as our ponies
picked their way carefully over these, the sound of
their hoofs striking against the rock rang hollow as if
caverns were beneath, which is probably the case.
Mostly, however, the spaces between the newer outbursts of lava have been filled in with sand-drifts in
stormy weather, and it is possible to cross the desert
by keeping to such places, skirting the newer lava-
beds which are utterly impassable on horseback. It
is toilsome work for the ponies as they sink fetlock
deep into the sand at every step; and one is compelled
to pursue a very circuitous route, skirting the rough

lava-beds until it is possible to ride round them. Fortunately, the beds appear to lie above channels radiating from *Askja*, and tend in the direction of the mountain; therefore in riding to and from that volcano one is not compelled to make such lengthy detours as would be necessary to cross from east to west. Both times that I have crossed we have headed as straight as possible for a gap in the highest part of the crater's mountainous periphery on the outward journey, and for the *Sellandafjall*, the westernmost of the mountains south of *Mývatn*, on the return. Only by the bearings of the mountains can the traveller direct his course, for compasses are totally unreliable by reason of the quantity of iron in the lava; therefore to be overtaken by a fog or a snowstorm, or worse still a sand-storm, when far out in the desert, would be a very serious matter, as under such circumstances it would be impossible to advance or return, and nothing could be done but await with all the patience at one's command for the fog to clear off, or for the storm's cessation. In the case of a fog or a sand-storm there would be no water, as only on the *Dýngjufjöll*, *Trölladýngja*, and the other mountains, or immediately at the foot of their slopes, is it possible for the traveller or his steeds to quench their thirst, the streams from the glaciers and snowfields being absorbed by the arid sand and porous lava at the very bases of the mountains.

It will be evident from the above description of the *Ódáðahraun* that it is not the most pleasant or safest region in the world through which to travel, especially as fogs, sand, and snow-storms of several days' duration are of frequent occurrence, even in summer, in this high latitude. Therefore, though it is only ten hours' ride from *Svartárkot* to the pass through the *Dyngjufjöll* into *Askja*, enterprising travellers who venture there are by no means sure when they will return; Professor Johnstrup's party when he visited the volcano in July, 1876, were snowed up in the crater by a snowstorm which lasted thirty-six hours, and made their escape therefrom with difficulty. Gunnlaugsson, as stated in the prefatory chapter, in the summer of 1837, when he made an excursion into the *Ódáðahraun* to map out that portion of the island, was compelled by fogs and snowstorms to abandon his project, and only with the utmost difficulty made his way out of the desert to a farm. Three years later, a German, Schythe by name, visited Iceland and made light of the perils attending travel in the interior, as Burton has done since, and, like the 'veteran traveller,' he failed ' to plant his lance in Iceland,' which was no less an undertaking than to prove that it was an easy matter during the summer months to cross the interior from west to east by the route taken by Gunnlaugsson, being driven back by a severe snowstorm with the loss, it is said, of several ponies. Therefore, although seldom a decade passed away without a volcanic eruption in this fire-blasted storm-swept wilderness, it is hardly to be wondered at that no one was ever bold or curious enough to visit the spots where these took place until February, 1875, when an exceptional

and adventurous Icelander, Jón, of *Víðikær*—of
whom more anon—crossed the *Ódáðahraun* from
Mýcatn to *Askja*, in the coldest month of an Icelandic
winter. There was another reason why the peasantry
were loath to venture into the desert, they believed
it to be peopled with a lawless race, the descendants of
men outlawed in the middle ages, and that the sheep
and ponies which frequently strayed away from the
outlying pastures on the borders of the *Ódáðahraun*,
and were never more seen, were carried off by them.
This summer, however, Jón, of *Víðikær*, induced
several of the farmers living in the north to ac-
company him on an exploring expedition, with the
object of ascertaining if there really was anyone living
in the lava-desert, or whether there were any pastures
to which the lost sheep and ponies strayed and there
lived in a wild state. I was at *Reykjahlíð* when Jón
and his companions returned, and he told me they
had passed ten days in the interior, and had
thoroughly explored the region south of *Mýcatn*
between the two rivers, to the very base of the
Vatna Jökull. They had found the skeletons of
many sheep and ponies which had strayed into the
dismal waste and there perished of starvation; and
also in a grassy oasis between the upper forks of the
Jökulsá, the remains of an old house, built wholly of
slabs of rock and lava without a particle of timber,
the roof of which had long since fallen in. The
bleached bones of sheep and ponies were lying
around; therefore, in all probability this house was
for some time the abode of outlaws, though possibly

at least a century ago. Thus we see that there was some foundation for the belief of the country people that the descendants of outlaws were living in the desert, and that their live stock were stolen. It is to be regretted that no attempt was made by the exploring party to clear away the *débris* of the roof to ascertain if the skeleton of the 'last man' remained in the house, where it is likely he perished unattended. This would have been no easy matter, Jón said, some of the slabs of rock that formerly roofed the house being so weighty that two men could not lift them.

Jón and I, unlike Gunnlaugsson, Watts, and Johnstrup, have been exceptionally fortunate as to weather, for during the whole time that Jón and his companions were absent the sky was without a cloud; and both in 1878 and this year the weather was all that could be desired during my excursions in the interior. Jón's exploring party were nearly caught in a sand storm, however, for within six hours after their arrival at *Reykjahlíð* it commenced to blow very hard, and we had a gale of wind that lasted four days, whirling up the sand from the deserts around so that one could not see two hundred yards from the house.

To resume the account of our journey. It was about seven o'clock when we quitted the *Sádrá*, and half-an-hour later we were amidst the lava of the *Ódáðahraun*. Through this for six weary hours we slowly pursued a zigzag route, heading as straight as possible for the chief of the *Dyngjufjöll*, which rises like a mountainous isle from the midst of a sea whose billows had been suddenly petrified into masses of rock. Being beautifully clear, the sun was visible until nearly eleven o'clock, when it sank behind the mountains in the N.N.W., and the peculiar twilight of an Arctic midsummer's night followed, the sky to the northward being illumed by a soft, subdued, mysterious light, while the landscape was wrapt in shadow. One's feelings at such a time, amidst surroundings literally indescribably weird, are as difficult to analyse as to express. So impressive was the scene that we rode along for several miles without exchanging a word, our ponies' hoofs the while falling noiselessly on the sand, some lines of Mrs. Hemans' beautiful poem, 'The Caravan in the Desert,' came vividly to mind:

> ' 'Tis silence all. The solemn scene
> Wears at each step a ruder mien;
> For giant rocks, at distance piled,
> Cast their deep shadows o'er the wild.
> Darkly they rise—what eye hath viewed
> The caverns of their solitude?'

The ascent of the mountain may be said to begin out in the desert at a point seven to eight miles distant from the highest part of the defile—named *Jónsskarð* (John's Pass) in honour of Jón of *Víðiker*, the first man to traverse it—in *Askja's* mountainous periphery, through which it is possible to descend into the crater. From here there is an upward slope, which appears to have been formed by a number of lava-flows having issued from the volcanic vent below *Askja* before its present encircling mountain-wall was builded up, and spread over the plain in a northerly direction, there being signs of a terrace-formation in places, as if each succeeding lava-flow had been less in bulk than the one preceding and had not extended so far. Owing to the time that has elapsed since the lava-flows issued, and the number of pumice and ash eruptions that have subsequently taken place from *Askja*, this terrace-formation is nowhere very distinct, save to the westward, where it slopes downward more abruptly than towards the north. The highest part of the slope thus formed has an altitude by Aneroid of about 3,300 feet, 500 less than that of the surface lava in *Askja*, and quite a thousand less than the highest point of the pass.

Ascending the slope in a south-easterly direction, on the left hand, for a distance of five miles, a number of crater-cones of quite recent formation, and hills of scoriæ, several hundred feet in height, show where even these vast superincumbent beds of rock have been fissured and upheaved by explosions within the mountain, when molten matters endeavoured to force a vent; while for a like distance on the right hand, there is a tremendous chasm, on the further side of which a mountain at least five miles in length from N.E. to S.W. rises precipitously to a height of 3,000 feet from its depths—4,000 feet above sea level. This mountain is not shown on any map of Iceland that I have seen, and, I think, must have escaped Lieut. Caroc's notice through hazy weather on his journey to and from *Askja*. Peculiar black domes, possibly of obsidian like 'Mount Paul' in the *Vatna*, cropped through the snow-covered summit of this mountain. On the map attached to this volume the eastern side of the mountain and the chasm are roughly traced.

From the last terrace the ascent becomes more steep, first over lava-flows that have forced their way through rifts in the defile, and later on over ice, and one must perforce dismount and drag his steed reluctantly after him. At 2 A.M. we had attained an altitude of nearly 4,000 feet, and, previous to rounding a projecting spur of the mountain which will shut out the Ódáðahraun from view, we turn to gaze over its gloomy wastes at the rising sun, which even at this early hour is returning from his brief sojourn beneath the horizon.

To our surprise, we discover that the desert is hidden beneath a thick stratum of mist which, propelled by a light northerly air, had unknown to us been following in our wake. To the northward, the mountains twenty miles distant encircling the Midge Lake on the south and east, rise through the mist like islands from a frosted-silver sea; and between two of them, a few degrees to the eastward of north, the sun is shining brightly from the centre of a lurid halo, while the sky above is of a beautiful pearly gray flecked with a few golden cloudlets. The mist, where it is tinted by the ray of light streaming between the mountains from the sun, gleams like a flood of molten metal, its brilliancy intensified by the far-stretching deep purple shadows cast by the mountains on either hand. As we gazed, by the stroke of Nature's magic wand, Heat, the whole scene began to change; first, the fiery flood of mist was rarified by the warmth of the sun's rays and slowly began to ascend, disclosing to view a broad stripe of the black lava below—the mist purple in the shadows of the mountains on each side remaining stationary. As the sun rose higher and higher, the whole of the mist was slowly set in motion, rising in wall-like masses radiating from the sun; and these, when they had attained an altitude above the air-current from the north, remained motionless, and at the moment when my companion inquired if I was going to stay where I was all day, formed three vast colonnades, between whose vapoury walls streamed rays of golden light, as if they led to some brightly illumed palace in the sky! while below, in strange contrast, mile after mile of the black fire-blasted wilderness was disclosed to view. It was an unique sight, suggestive of Chaos being reduced to order by command of the Creator—of the moment when the fiat went forth, ' Let the waters under the heaven be gathered together into one place, and let the dry land appear.'

Resuming the ascent we rounded the spur of the mountain, and entered the defile. Its direction is from N.N.W. to S.S.E.; and it is about half a mile in width throughout its whole length, which I should judge slightly exceeds two miles; a steep ice-covered declivity between two mountain walls whose jagged peaks rise on either hand to a height of nearly a thousand feet. Several small crater-cones built up of a peculiar bright-red lava-slag crop through the ice, extensive tracts of which were quite black in colour owing to the immense quantities of volcanic ash embedded in it. There was not nearly so much ice in the pass this year as in 1878, the present and preceding summers having been unusually hot, but what there was was about forty feet thick, very rotten and dangerous, and full of crevasses, which enabled us to perceive, as we dragged our ponies reluctantly by their bridles after us up the slippery incline, that we were passing over the roofs of icy caverns, and that a considerable stream of water was coursing impetuously through them over a rocky boulder strewn bed lying in some places fifty to sixty feet below.

East of the pass there is a grand Alpine valley filled with a glacier at least three miles in length, and half a mile in breadth. In 1878 Einar, instead of heading south through the pass, turned to the left and took me for a couple of miles over the glacier in this valley before he discovered that he was going in the wrong direction. This mistake is not likely to be made by future visitors who may have other guides than Einar, if it is borne in mind that the highest part of the pass is marked with a peculiar rocky pinnacle, visible from where we turned astray. This pinnacle stands quite isolated, slightly to the right of the middle of the pass, and rests upon a low rocky ridge that intersects the ice at right angles to the acclivities on either hand.

By 3 a.m. we were abreast of this landmark; and a few hundred yards further brought us to the southern end of the pass, where it terminates abruptly in a precipitous declivity of ice and ashes. From the verge of the precipice a splendid view over *Askja's* weird amphitheatre is obtainable. Notwithstanding that this was the second time I had looked upon this terrible arena where for countless ages two powerful gladiators, Fire and Frost, have been struggling for the mastery, I could have sat upon my steed and gazed for hours upon the scene. The huge crater lay wrapt in deep shadow some eight hundred feet below, the highest peaks of its mountainous periphery on the farther side alone as yet illumed by the rising sun. It required no great stretch of one's imagination to fancy the two gaps nearly opposite each other in the crater's encircling wall openings for the admission of the combatants, and that a brightly-illumed cloudlet which at this instant streamed through the one in the north-east was the chariot of the Fire King, who was hastening to renew the conflict which had been taking place here at intervals for ages, and that the vast columns of steam ascending in the south-east part of the crater marked where his antagonist awaited his coming upon the very spot where the last struggle had taken place.

Having gazed for some minutes at the indescribably weird scene that *Askja* presents when seen from the pass, we prepared to descend into its depths. My former experience had taught me that this was no easy matter, for the descent is first down a steep icy declivity for three or four hundred feet, and then for a like distance down a steep slope of loose ashes. Leading our ponies we made our way without much difficulty down the ice, to discover, however, that it terminated in a sheer precipice thirty feet in depth. We were completely at a nonplus; but fortunately after a short search we found a breach in this, and were able to scramble down to the ashy slope below. I had a narrow escape from being crushed under my pony during this part of the descent, for he lost his footing, and as I was going first leading him by the bridle, he fell heavily upon me, and we both slid down the ice until we brought up in a bank of ashes. It is almost unnecessary to observe that I drove the brute before me during the remainder of our descent. The ashy slope being very steep, it abounded in yawning rifts where 'slips' had taken place, accordingly our progress at times was far more rapid than safe or pleasant; our poor steeds in their descents into the rifts being several times nearly buried beneath the ashes that accompanied them. Only Icelandic ponies, which it has been truly said will safely traverse places that would shock the nerves of a goat, could ever have been led down such a precipice.

Fortunately we succeeded without any more serious mishap in finally reaching the floor of *Askja* : we had not done so badly, having accomplished the journey from *Svartákot* in eleven hours.

Before proceeding to cross its lava-covered floor, it will perhaps be as well to more fully describe this vast crater than I have hitherto done. It is almost circular in shape, quite seventeen miles in circumference, and encircled by a somewhat jagged mountain wall varying from 800 to 1,500 feet in height above the superficial lava in the crater. This mountainous periphery is highest on the south and north, and lowest in the north-east, where it does not rise more than 800 feet above the floor of the crater for over a mile; at least three of its highest hollows contain true glaciers, and its peaks are snow-clad ten months out of the twelve. East of the lowest part of the wall there is a gap to the level of the surface-lava in *Askja*, through which lava has coursed down the outer slope, and spread over the *Ódáðahraun*. There is also another gap almost directly opposite in the south-west, but whether lava has ever found an outlet there I am unable to say, not having been in that part of the crater. Mr. Watts is the only person who has been through this gap, and, as before mentioned, he says he descended 'over a lava-stream which here enters from the Ódádahraun, and had run for some distance uphill.' From the appearance of the gap in the north-east, and the altitude of the lava deposits in *Askja* that have issued there, 2,300 feet *above* the *Ódáðahraun*, I am inclined to the belief that Mr. Watts must have been deceived by some peculiarity in the appearance of a lava-flood *from* the south-west gap, or that his 'vision distorted by fatigue and sleepiness' saw things as in a mirage reversed. It is possible that my surmise is correct, and that a lava-flood flowed out of the south-west gap in the following manner, and now presents the appearance of one that had 'run a considerable distance uphill.' It issued in the winter season, and its outer surface congealing rapidly, a covered way was formed, through which the molten flood continued to course downwards until the supply ceased for a short period, when the tail end of the flood congealed, and formed a core within the covered way by which the lava was stopped upon the flow recommencing, and this remains piled up in *Askja*, and now appears as if it had flowed—I had a good mind to have penned flown—upwards from the plain. If my memory does not play me false, I saw a small lava-flood as above described among the volcanic mountains east of *Mývatn*.

Professor Johnstrup states that the lava on the eastern side of *Askja* has a declination towards the gap in the north-east of 300 Danish feet in a stretch of 12,000 = to 1 in 40.

Now, as at the time of my former visit, the floor of *Askja*, notwithstanding it lies 3,700 to 3,800 feet above sea-level, was bare of snow. This, one would imagine, must be owing to internal heat, as twenty-five miles to the southward we have the glacier-covered region known as the *Vatna Jökull*, with a mean altitude of less than 5,000 feet; and in the north-west of Iceland, at an altitude of less than 3,000 feet, the icy-wastes of the *Glámu* and *Dránga Jöklar*, which tracts cannot possibly be more favourably formed for glacial deposit than *Askja's* amphitheatre. Professor Johnstrup says that in Iceland the snow line can be set at about 2,500 feet (Danish). The whole surface of *Askja*, save in the south-east where there is an extensive tract covered with pumice erupted in 1875, and a tepid lake five miles in circumference, is a chaos of rugged lava-floods that have issued here at different periods. From those on the left, looking south across the crater, for an area exceeding a square mile, ascend innumerable small jets of steam; and there is a solfatara of small extent between this tract and the mountain-wall on the north. These do not mark a site of disturbance during the 1875 eruptions, however, the situation of the rifts and vents then opened being clearly indicated by the enormous volumes of steam that belch forth on the further side of the crater, close under its encircling mountain-wall in the south-east.

The five miles of rugged lava and pumice between us and the vents opened in 1875, I knew only too well, would have to be crossed on foot; therefore, after partaking of some food, H., Arni, and I started, leaving Einar in charge of the ponies. In 1878 we hobbled our animals, and left them unattended at the foot of the pass, to discover upon our return that two of them had broken their hobbles and made tracks for the *Súðrá*, and that the third was anchored fast half-way up the ashy slope by a large stone to which, being a very valuable animal, he was fastened by a lariat for greater security: consequently, Einar, I, and another man—the Deputy Sheriff, by-the-bye, and a first-class fellow—after being a day and night out of bed had to return across the *Ódáðahraun* on foot, the remaining pony being laden with the saddles of the others. How ever I accomplished the journey I do not know to this day, for I was delirious when I reached Einar's roof.

The crossing of the lava-covered floor of *Askja* is most fatiguing work; it has taken me, a young and active man, each time that I have crossed, four hours to proceed as many miles, most of the way by the aid of my hands protected by thick woollen mittens that they might not be cut by the lava. I may also observe for the benefit of anyone who in the future may visit this volcano, that the crossing of *Askja* utterly ruins the pair of boots worn, the sharp edges of the lava cutting through the sides of the uppers like knives, so that an old pair with good but not thick soles should be taken, as their destruction would not prove so great a loss as that of a newer pair.

The superficial lava in *Askja* is pronounced by Professor Johnstrup to be basaltic; and it has evidently issued at different periods, some tracts being covered with a lichen growth, while others are as rugged, black, and new-looking as those lava-floods near *Mývatn* which issued but a century-and-a-half ago. I saw no lava anywhere so new-looking that it could possibly have been erupted as recently as 1875.

When one has succeeded, after many rests and tumbles, and much scrambling, in approaching within a mile or so of the bursts of steam, he is able to walk upright, and proceed more quickly; the lava being buried beneath a covering of pumice which gradually increases in depth as the vents whence it was erupted are approached. The pumice is of three colours: black, light-silvery-gray, and golden-brown; the last-named very fibrous and presenting the appearance of masses of the outer-husks of gigantic cocoa-nuts, and when blocks of the two last-named were broken, the newly exposed surfaces respectively glistened like silver and gold. This substance must have been much harder and tougher when ejected than now, for the blocks, some of which were three to four cubic-feet in bulk, when lifted breast high and let fall, not cast to the ground, broke into fragments by their own weight. Yet the blocks experimented upon must have been cast to and fallen from a considerable height, as they lay half-a-mile from the nearest of the newly-formed craters. The pumice, owing to the expansion by frost of the snow and rain which find their way into its pores, is fast degrading into a pumiceous sand. Amidst the pumice are a number of immense blocks of obsidian and pitchstone, some of which must be several tons in weight. These had evidently been erupted at the same time as the pumice, or subsequently.

The apex of the slope thus formed is a cone-shaped crater, about 200 feet in height above the superficial lava in *Askja*. When I was here in 1878, tremendous blasts of steam were belching forth almost continuously with perfectly deafening roars, a hundred times louder than those concomitant to the blowing off of steam from the boilers of the largest of Transatlantic steamships; but now, to my astonishment, all was still, and upon climbing to the summit I found that in the crater, at the depth of about 150 feet, was a placid pool of apparently cold water, with a number of small stufa emitting inconsiderable jets of steam within ten feet of its surface. The diameter of the crater at its mouth is at least 500 feet, and the interior is an inverted cone-shaped hollow, which decreases to one-third of that diameter at the level of the water. In 1878, there being no water in the crater, I could see that at about a depth of 200 feet a flat shelf, presumably the lava-strata in *Askja*, twenty feet in width, encircled a well-like opening in the centre, through which the steam belched forth. The walls of the crater are formed entirely of pumiceous sand and a clayey loam. I saw no traces of scoriæ, or lava-like slag. This crater is beyond a doubt 'the shaft like the mouth of a large coal-pit,' mentioned by Mr. Watts, though it is not 'situated in the N.N.E. corner,' but in the south-eastern part of *Askja*, there being no other vent here that will answer Mr. Watts' description. This crater in its present condition is a twin-brother to the large one on the western slope of the volcano *Krafla*, known as *Helvíta stœrra* (Greater Hell).

By Aneroid (in 1878) the summit of this crater has an altitude of close upon 3,800 feet. Taking the depth of the pumice at its base at 100 feet, and the height of the outer crater walls at about the same, the superficial lava in *Askja* at this spot lies 3,600 feet above sea level. Thus the altitude by my Aneroid agrees very closely with that marked upon Lieut. Caroc's map, viz., 3,660 Dan. feet=3,768 English.

The view from the summit of this crater-cone within a crater is as unquestionably unique as it is wildly weird; and one can form some idea of the terrible nature of the explosion that caused the earthquake on the memorable 4th January, 1875, from the veritable chaos of hugh masses of rock then disrupted and upheaved. The very slopes of the crater-cone, on whose summit we stand, rise on the south-west, at angle of about 50° to a height of 600 feet from the surface of a tepid lake, whose bed was formed, as before stated, by the disruption and subsidence bodily into the abyss beneath *Askja*, of an enormous mass of the lava deposits lying in strata above it, oval in shape and five miles in circumference. From our coign of vantage, *Askja's* mountainous periphery can be seen to greater perfection even than from Jóns Pass: on the south, distant but a little over a mile, the highest portion of it rises precipitously from the water's edge to a height of close upon 2,000 feet, while on the west, from the slopes of the crater round the north side of the lake, a distance of over two miles, the cliffs where the vast mass broke away rise a sheer semi-circular wall, 400 feet in height from the water to the level of *Askja*. This wall shows the lava deposits that have filled *Askja's* huge crater to its present level, *en profile*; and the stratification is so perfectly marked by the layer of scoriæ that has formed the surface of each deposit that it looks like a gigantic piece of masonry, fresh from the hands of Titanic masons; but this was more noticeable two years ago than now, large masses having since then broken away in places; and still more so in 1876, when seen by Professor Johnstrup, who says that the fresh appearance of the face of the rocky walls from which the detached mass broke away is proof that the subsidence of the north-westerly portion *at least* happened as recently as during the 1875 eruptions. There can hardly be any doubt that the whole was disrupted at that time, not the north part only, for, seen from the cone-shaped crater through a powerful field-glass, the encircling cliff on the south-west, at its junction with *Askja's* mountain wall, appears as new-looking as farther north.

East of the lake, and distant from it about 700 yards, *Askja's* encircling mountainous periphery is a vertical precipice, at least 800 feet in height, for nearly a mile; and it is evident by the face of this and the nature of the slope therefrom to the lake, that the slope is a huge slice of the mountain that at some period during the eruption had been blown bodily inwards. The explosion which hurled aside this immense mass opened a rift at least 200 feet within the mountain, the course of the rift being clearly traceable, though filled with débris, by a line of fumaroles, around which I noticed a trifling efflorescence of sulphur. This was the only place on the site of the 1875 eruptions where I noticed any signs of this mineral. Branching off the southern end of the rift, but divided from it by an angular rocky ridge, a deep gorge runs in a south-easterly direction for some distance into the encircling mountain, and here the immense volumes of steam, seen from the further side of the crater, were roaring forth.

From the above the reader must try to form some idea of what the scene of the 1875 eruptions in *Askja* is like, for I must confess I am unable to give anything like an adequate description, and I doubt if even the pen of a Dante could do so, though, perhaps, the pencil of a Doré might.

My companion, H., whose respiratory organs are

none of the best, now declined to go farther, therefore Arni and I continued our explorations alone. With considerable difficulty we made our way over the rocky débris, lying in the rift, to the gorge in the south-east. Climbing to the summit of the rocky ridge, bordering its mouth on the north, that we might command as good a view of it as possible, we could see, when the wind for a moment wafted the clouds of steam away, that the gorge extended in a south-easterly direction from the waters of the lake into the mountain for close upon a thousand yards, and that its declivities were too precipitous for a descent to be practicable without a long rope, with which, unfortunately, we were unprovided. I endeavoured, but fruitlessly, to borrow one at *Svartákot* before setting out for the volcano; and I now greatly regretted that I had not brought one from England, there being reason to believe that very large Geysirs exist in this part of *Askja*. The bottom of the gorge lay about 600 feet below the spot where we stood, and 50 above the surface of the lake, from which it slopes gradually upwards. Huge volumes of steam—I feel almost certain, likewise hot-water—belched forth at intervals from about a dozen large holes, or rifts, with such violence that the rocky ridge upon which we stood trembled. There were also a large number of small jets of steam jetting out horizontally from holes in the precipice opposite. Owing to the vapoury clouds that accompanied the more noisy outbursts we were unable to see whether any water was ejected or not;

however, I think it almost certain such was the case, and that one or more Geysirs here exist, the stream of steaming hot water that was flowing down to the lake being too considerable to have been formed by condensed steam alone. By the violence with which the steam here escapes it is evident that even at this time, five years after the eruption, the heat in the volcanic vent, of which *Askja* is the outlet, must be intense, and the pressure of steam very considerable, and that were it not for the vents, indeed, safety-valves would be the fitter term, which here exist, another terrible eruption might at any moment take place.

While passing along the rift I noticed, projecting through the débris, some immense masses of obsidian and pitchstone that appeared to have 'boiled' up towards the conclusion of the eruption; therefore I resolved to return the same way and secure some specimens. I was fortunate enough to obtain here one piece of obsidian that throws some light upon the formation of the black pumice, it having apparently changed into that substance upon one side, where it had been subjected to intense heat.

Professor Johnstrup expresses an opinion that the whole of the pumice and ashes were ejected from the cone-shaped crater north of the rift, which he accordingly christened the *Pimpstens Krater*—*i.e.*, Pumice Crater. I am, however, of a different opinion: I believe that from the rift throughout its whole length pumice, obsidian, pitchstone, and ashes were ejected as well as from the crater, the latter marking the termination of the rift, and being builded up into its present cone-shaped form towards the conclusion of the eruption, when the rift was blocked up with the débris of the side of the mountain. I think that this is proved by the presence of the blocks of obsidian and pitchstone scattered about among the pumice, for had those substances been ejected before the last named, they would have been buried under it, and if subsequently, they would have been resting on, not deeply inbedded in it.

It is recorded that the greatest eruption of pumice and ashes took place on the morning of the 29th March, 1875, the eruption being preceded by a sharp shock of earthquake; but according to the letter before alluded to, written by a resident in *Reykjavík*, and dated the 23rd of that month, a considerable eruption must have taken place prior to that date, for the letter says: 'Ashes, too, had fallen over the north-east coast, so that pasture fields were covered so that the farmers had to take their sheep into the huts and feed them.' Whenever the eruptions took place, it is certain that strong westerly winds must have prevailed, for but a comparatively small quantity of the pumice ejected fell in *Askja* west and north of the rift and crater, the bulk being borne away to the eastward and scattered over the country in that direction, vast quantities being carried out to sea, as mentioned in the prefatory chapter. Professor Johnstrup states that over 3,000 square miles of country east of the volcano, fortunately chiefly sandy wastes and barren moorland, were covered with pumice by this eruption, the minimum depth near the coast being about two inches, the depth gradually increasing as the volcano was approached. The pasture-land belonging to several farms east of the *Jökulsá* was rendered useless for some years to come, and pecuniary assistance was sent from England to the farmers ruined in consequence.

Askja, it is believed, was visited but twice during the eruption—first in February, by Jón of *Víðikær*, and later on, in July, by Mr. Watts. As the latter's account of what was taking place at the time of his visit throws some light upon the formation of the bed of the lake, I beg permission to quote a couple of paragraphs. Alluding to this abyss, which he erroneously, it would appear, terms the crater, Mr. Watts says (p. 86 ' Across the Vatna Jökull ') : ' Presently, apparently about a mile away to the north, we could see the rim of the crater, at a great depth beneath us (Mr. Watts was on the summit of the mountain south of *Askja*), and while we were looking at it, a great crack opened upon the margin, and a huge slice slipped with but little noise into the crater—deep, deep down beyond the range of vision.' From this it is reasonable to believe that the disrupted mass did not sink at once to its present level, but subsided slowly, possibly continuing to do so as long as lava streamed forth in the *Mývatn's Öræfi*. There can be no mistake as to this being the ' crater ' of Mr. Watts, notwithstanding that it is oval, not ' triangular in shape,' for he mentions its circumference correctly— viz., five miles—and it is the only abyss in *Askja* of that magnitude; but with reference to its being a crater, Professor Johnstrup states, with the sentence emphasised in italics, '*that there was no evidence that the site of the lake had been a crater.*'

The 'Pumice Crater' north of the rift at this time, it appears, did not emit steam, but smoke, for Mr. Watts says further on: 'A shaft, like the mouth of a huge coal-pit, was disclosed to the N.N.E. corner of the valley, but beyond the rim of the crater, from which a column of pitch-black vapour was issuing— boom! boom! from its hoarse black throat, was succeeded in a few seconds by a heavy shower of coarse, earthy granules. . . . Suddenly a fearful crash made us stand aghast; it seemed as if half the mountain had tumbled in upon the other side of this horrible valley . . . and huge wide cracks, even where we stood, showed us that our position was not altogether a safe one.' Under such circumstances exploration must have been next to impossible, and therefore the imperfect description of *Askja*, both in Mr. Watts' paper read before the Royal Geographical Society, and in his book, is not to be wondered at, though the strange hallucinations he labours under as to the size and situation of the volcano are, for on p. 189 of the book quoted he says 'Öskjugjá can only be regarded as a lateral crater of the Vatna'! whereas it is distant therefrom at least twenty-five miles.

Respecting the lava deposits in *Askja*, bared by the subsidence now of the bed of the lake, I beg leave to translate an interesting paragraph from the paper by Professor Johnstrup before alluded to:—

'An excellent insight into the history of *Askia's* formation is here afforded, the vertical fractured surfaces showing *what a multitude of lava-floods must have been deposited in Askia's cauldron shaped valley (kjedelformige Dal)*, one above the other. (Italics *sic*.) The divisions between these lava-floods are distinctly marked by the layers of red slag-like lava, which time after time has formed the surface of the underlying lava strata; and I doubt very much if there can be found in any other place in Iceland, except the Almannagjá, where, however, the formation is far from being so distinct, such an instructive and grand profile as this. It has more than ordinary interest, owing to the striking resemblance presented by the volcanic deposits here to those widely spread rock-formations (*Bjergdannelser*) of basalt and dolerite which have been pronounced by most geologists to be of plutonic origin (*plutonisk Oprindelse*). Had they had an opportunity of viewing this profile, they would certainly have entertained a different opinion.'

When the Professor visited the volcano, he says the surface of the disrupted mass lay 740 Dan. feet lower than its original level; and according to Lieut. Caroc's map a lake existed in the south-eastern part of this, nearly circular in form, and 4,400 Dan. feet in diameter, its water lying at an altitude of 2,885 Dan. feet above sea level. Watts makes no mention of any lake, therefore in all likelihood none existed at the time of his visit, the one seen by Caroc being formed by the hot water streaming from the gorge, and the surface water in *Askja* draining into the abyss during the twelve months that had elapsed since Mr. Watts was there. In 1878 I found that in the two years' interval since the Professor's visit the lake had greatly increased in size, its water then covering the whole surface of the subsidence to a considerable depth; and from the level of the water this year (1880) I should say that it had risen quite forty feet since 1878. This is only what was to be expected, as in all probability the greater part of the snow and rain which fall over an area of nearly twenty square miles drain through radiating rifts into the abyss. The Professor states that he found the temperature of the water 22° Celsius = 104° Fah., but when I tested it in 1878 it was only 97° Fah. Having, unfortunately, in the hurry of landing my luggage at *Akureyri* forgot both my Aneroid and my thermometer, I was unable to ascertain the temperature of the water this year, nor with exactitude how much it had risen since I was here last.

The quantity of water must continue to increase, there being no outlet, and in the course of some twenty years, should no eruption take place in the meantime, and open fissures to admit the water now collected into the heated depths below, the basin formed by the subsidence will be completely filled. This large and ever increasing lake in the bosom of an active volcano is a most alarming feature; it is extremely likely that when the water rises high enough to invade the gorge that it will find its way below and cause an explosion that will let the whole contents of the lake come in contact with the molten matter that it is reasonable to believe will lie at no great depth, a terribly violent explosion must inevitably ensue, one that will be likely to cause an earthquake to which those of 1872 and 1875—which opened rifts, it will be remembered, thirty miles in length at a considerable distance from the volcano—will be comparatively insignificant, one that will in all probability not only split the mountain into sections to the level of the *Ódáðahraun*, and thus make Watts' description tolerably correct, but also send back the tide of molten matter in the channel connected with the earth's molten interior underlying Europe, in a way that may cause considerable volcanic disturbance on the Continent, there being reason to believe, as before observed, that such a channel does exist, and is connected with Iceland, the great Lisbon earthquakes in 1755 being preceded by the commencement of terrible eruptions from the *Kötlugjá*, which lasted a year, while the Calabrian earthquakes, thirty-two years later, were followed by the outbreak of the prodigious lava-floods in the vicinity of the *Skaptár Jökull*.

Being pretty well tired out I now rejoined H., and lay down on the slope of the 'Pumice Crater' for a short time, to recruit and partake of a little food before setting out to recross the lava-covered amphitheatre. During our rest we watched with considerable interest the gyrations of a pillar of sand about 200 feet in height whirled up by the rotary motion of the winds in *Askja*. This pillar travelled all over the crater in a most remarkable manner—now coming menacingly towards us, as if it would overwhelm the daring invaders of a spot where earthquakes have their birthplace, and anon retreating towards the south-west gap, as though beckoning us to retreat at once in that direction before harm befell us. This phenomenon is easily accounted for. In the north and east of Iceland, during the summer months, sea breezes prevail, which gradually increase in strength from very early morn till four or five o'clock in the afternoon, after which they gradually subside. These sea breezes are due to the air above the *Ódáðahraun*, rarified by the heat of the sun refracted from the lava, constantly ascending during the day-time, the cold air-currents from the sea rushing in to take its place. To-day these winds appeared to meet in *Askja* another air-current coming from the icy wastes of the *Vatna Jökull* in the south, and formed a whirlwind which whirled up the light pumiceous sand and fine ash in the manner described. While resting I made a couple of sketches of the crater, one looking towards the western gap, in which I included this pillar of sand, and from these, since I came home, I have made a couple of water-colour drawings, which convey a very fair idea of *Askja* as seen at this time.

Somewhat refreshed with our rest we now set out to recross the crater. How we each longed for a pair of wings to bear us over the four miles of rugged lava that intervened between us and Einar and our steeds! but, alas! longing availed nought! and though we were in a spot weird enough to be the home of Genii, none appeared with the wished for wings. Shortly after we had passed the pumice, deluded by a somewhat level tract of lava, we took a more westerly direction than we had taken on our way over to the site of the 1875 eruptions. This led us into a labyrinth of fissures, and finally brought us face to face with an extensive bed of exceedingly-rugged and comparatively new-looking lava, which had welled up through a rift but a century or so ago. This was the newest lava I saw in *Askja*, though I saw some newer-looking on the outer slope of the mountain east of the entrance to the pass; therefore, probably, it is the last which issued here. We debated for some little time whether we should skirt or cross it, finally deciding to do the latter. We found it terrible hard work, and were compelled to sit down and rest every few yards or so. When I started from the 'Pumice Crater' I had my pockets laden with about a quarter of a hundred-weight of fragments of various kinds of rocks; but every time that I sat down I examined these to see if there were any not worth the trouble of carrying farther; some of the pieces I handed to Arni to carry, who took them very reluctantly, the bulk, however, were by degrees thrown away, each succeeding rest seeing one or more of the specimens abandoned which at the rest preceding I had determined to preserve; greatly to the amusement of H., who is not disposed to subject himself to the least inconvenience for the cause of science. When we ultimately reached the spot where we had left Einar, Arni and I had not a fourth of the quantity with which I started. It took us five hours to reach him; and I shall never forget poor H., who was the last to arrive, he was completely beaten, and it took him at least half-an-hour to cross the last two hundred yards of lava.

After a short rest we led our ponies up the ashy slope to the glacier-filled pass, which we traversed in safety; and fourteen hours later we were safe once more under the roof of *Svartákot*, having been absent forty-two hours.

CHAPTER IV.

THE PROBABLE GENESIS OF ASKJA AND ICELAND.

> 'The earthquake then,
> As a beast in pain,
> In his burning den,
> Snapt his chain :
> Till bound at last, like a snake he curled,
> And formed the mountains of our world.'
>
> (JOHN MILL.)

N this chapter I purpose bringing this monograph to a conclusion with a few observations upon the genesis and probable history of *Askja*, and likewise upon the part played by that volcano in the formation of Iceland; but before doing so, I will, now that my reader has figuratively accompanied me over both the scenes of eruption in 1875, give my reasons for asserting so positively that the earthquake which opened the rifts in the *Mývatn's Öræfi* was caused by an explosion in *Askja*, and that the lava came thence through a subterraneous channel. Fire, smoke, and steam were seen ascending from *Askja* immediately after the earthquake on the memorable 4th January, and on the night of the 9th the column of fire was so high and brilliant, that according to a letter written

by a resident of the capital—which is quite a hundred miles distant: 'all agreed that it was some neighbouring farm burning, with haystacks. The fire shot up like lightning, . . . when daylight dawned, and we could discern the mountains, we observed a thick and heavy column of vapour or steam far in the background beyond all the mountains visible. . . Morning and night this grand display was visible during the 9th, 10th, 11th, and 12th, and during the day the column of steam and smoke stood high in the sky.' During the whole time that lava was flowing forth in the *Mýratn's Öræfi* smoke and steam were seen ascending from *Askja;* and prior to the outburst of lava being discovered, explosion after explosion took place there with such violence and frequency that Jón of *Viðikær*, as before related, ventured to cross the great central lava-desert, in the coldest month of an Icelandic winter, to see what was taking place. Moreover, while no lava was erupted from *Askja* itself, the quantity of pumice there ejected was prodigious, and proportionate to the magnitude of the lava-flood that issued in the *Öræfi*, where only an infinitesimal quantity of pumice and ash was thrown out; and these facts, taken conjointly, are almost irrefragable proof that the lava came thence, for pumice is asserted by mineralogists to be the scum which forms on the surface of lava when in a molten state.

The beds of basaltic and doleritic lavas underlying *Askja*, bared in the face of the cliffs bordering the tepid lake on the north, are conclusive evidence that a huge cauldron-shaped hollow formerly existed in the heart of the chief of the *Dýngjufjöll*, which has been filled up to its present level by the lava-floods that have been deposited therein one above the other. It is reasonable to believe that in the course of time these vast deposits first narrowed and then entirely blocked up the vent or vents whence the lava issued, so that in 1875 it required less force to upheave, at a lower level by some 2,400 feet, the rocky roof of an old channel * running under the *Öræfi*, connected with the volcanic outlet amidst the *Dýngjufjöll*, than to force a vent through the lava deposits in *Askja* itself. To judge from what occurred at both places, it is probable that the force required to burst asunder the strata forming the roof of the channel in the *Öræfi*, and that necessary to force an outlet in the crater, were nearly equal, and that at the same moment the rocky roof of the channel lifted in the desert, the huge oval mass that has sunk in *Askja* was also forced from its bed. Naturally, vents having been formed in two places, the heavier molten matter would flow from those at the lower level, while the steam and gases would make their escape, carrying with them

---

* The course of this channel is, in all likelihood, marked by the depression in the *Mýratus Öræfi*, running northward from the 1875 lava-flood, bordered on the east by an old rift known as the *Sveinagjá*, as at a spot where a line continued from it in a north-north-easterly direction would intersect the bed of the *Jökulsá*, the river bordering the desert on the east, there is a group of small new-looking crater-cones, and the surface of the earth has been greatly disturbed by recent earthquakes. The *Sveinagjá* runs some distance southward beyond the northern end of the 1875 lava-bed; and the subsidence there took place between this older rift and a newly-formed one.

the lighter substances, such as pumice and ashes, from the vents at the greater altitude. The latter, moreover, would act in a manner as safety valves, and the molten lava, not being subjected to any great pressure from confined steam or gases, by its own gravity would well gently forth through the newly-formed vents at the lower level, instead of being forcibly hurled far and wide by the enormous pressure of confined steam and gases generated above it, as most certainly would have happened if the 'safety valves' in *Askja* had not existed.

Fortunately for the inhabitants of Iceland this is exactly what occurred, as we have seen.

The shock of earthquake on the 15th August, and the slight eruption of ashes and volcanic bombs in the *Öræfi* from the crater-cones at the northern end of the lava bed are, I think, easily accounted for. It is extremely likely that a quantity of water had been gradually collecting in the hollow in *Askja*, now the bed of the lake, and that at this time it suddenly found its way through a newly-opened fissure into the abyss beneath (that one exists the subsidence conclusively proves), in which the molten lava had then fallen so low that its surface lay lower than the level of the roof of the subterranean channel connecting *Askja* with the scene of eruption in the *Öræfi*, that the steam generated caused the shock of earthquake, and forced out the pumice and other substances forming a scum on the lava in the channel. The fact that lava had ceased to issue for some time previous to this shock of earthquake confirms the view taken, that the channel was only partly filled with lava, so that a scum could form upon that lying therein ; and from what was taking place in *Askja* a month earlier, when Mr. Watts was there, it is not very speculative to imagine that a fissure was newly opened, and that through it the water from the snow, rapidly melting under the summer sun, found its way into the heated abyss.

It will doubtless interest my readers to know what Professor Johnstrup, in his paper before referred to, says with reference to the connection of the two scenes of eruption. 'The volcanoes here (the *Öræfi*) and in Askia stand in far closer connection with each other,' the Professor says, 'than one would really expect, considering the distance they are apart, and the great difference in the volcanic operations at the two places, and also in the ejected substances.' But he appears to have been deterred from actually stating that he believed the two scenes of eruption to be connected, by the importance he attaches to the facts that 'two volcanoes near each other, and both active at one time, should be so opposite in character and should eject such totally different substances as pumice rich in silica, and basaltic lava so poor in that respect.'

The Professor appears to utterly ignore the fact, that the difference in the substances ejected is almost in itself conclusive evidence of the connection of the two scenes of outbreak, for surely the paucity of silica in the lava that welled forth in the *Öræfi* is fully accounted for by the tremendous pumice eruptions from *Askja* in January, and on the 29th March! Still these peculiarities in the 1875 eruptions are very interesting facts, as they confirm the opinion of mineralogists that much of the perplexing variety of textures of lavas arises from the lesser or greater intensity of the heat, and likewise of pressure from gases to which they are subjected when in a molten state.

It cannot be questioned that the chief of the *Dýngjufjöll*, with its vast crater *Askja*, is justly entitled to rank as 'Iceland's Largest Volcano,' notwithstanding that no earlier eruption there than those of 1875 is recorded, and that prior to February of that year it was not known that an almost circular crater, having an area of at least twenty-three square miles, was embosomed in the largest mountain-mass rising amidst the *Ódáðahraun*. It may be that its highest peaks, those south-west of the lake, do not attain so great an altitude by three or four hundred feet as that said to be attained by the summit of the *Öræfa Jökull*, 5,927 feet (Watts), but it is also possible that the altitude of that volcanic *Jökull* has been exaggerated, and I think it likely that when the altitudes of both mountains have been ascertained with more nicety than at present, it will be found that the culminating point of the island is the chief of the *Dýngjufjöll*, whose vast outer circumference causes it to appear, when seen from a distance, a far lower mountain than it really is.

It is not unlikely that the early settlers, or their immediate descendants, were aware that this mountain was a volcano. Those 'hardy Norsemen' who sought a home in Iceland rather than abandon the Pagan religion of their forefathers at the behest of a king, whose

> Banners braved the gales of the western world,
> Long ere Columbus that of Spain unfurl'd,

are not likely to have left their island home unexplored, and there can be no doubt *Askja* was visited. This is conclusively proved, I think, by the name borne from time immemorial by the mountains around, Bower-mountains, the bower being the crater *Askja*. In the course of time the verbal accounts of these early explorations became forgotten lore; and, as nothing was to be gained by venturing into a wilderness of igneous rock, the interior of Iceland east of the *Sprengisandr* (Bursting Sand) became a *terra incognita* to the listless and apathetic, yet kindhearted and hospitable modern Icelanders, till Jón of *Víðikær*, in 1875, and Mr. Watts, in 1876, showed that the Misdeed-lava-desert is not so impassable as was believed.

Although seldom a decade passed away without a volcanic eruption in the *Ódáðahraun*, no one was ever bold or curious enough to visit the spots where these took place until Jón did so in 1875; and it was but seldom the trouble was taken even to record the date or duration of an eruption. Volcanic eruptions in Iceland more often than not take place in thick weather, and, moreover, are invariably attended with the emission of great quantities of steam, accordingly from the inhabited coastal districts (I believe that in the whole island there are not more than half-a-dozen

inhabited houses distant in an air-line over forty miles from salt-water) it is impossible to see, if the eruption is any great distance off, the exact spot where it is taking place. We had an example of this in 1875. Burton publishes (p. 46, vol. 1) the letter from a *Reykjavik* correspondent before quoted, and as it illustrates the careless way in which the Icelanders record volcanic eruptions, and 'jump to conclusions' as to the volcano active, I will give another extract therefrom, first quoting Burton's prefatory remark :
'The year after the author's departure witnessed an eruption of the Skaptárjökull, in the north-west corner of the Vatnajökull, but it lasted only four to five days. The following account appeared in the papers; nothing more has subsequently been learned about it.'

" On Thursday the 9th January, about three o'clock A.M., we observed from Reykjavík a grand fire in east-north-east direction. (The centre of the *Jökull*, credited with the eruption, I would here observe, lies due east of *Reykjavík*, while the bearing of *Askja* is as nearly E.N.E. as possible, being 24° north of east.) But when daylight dawned...it was clear that it was far off, and, according to the direction, it seemed most likely to be in Skaptárjökull...When similar news came from east, north, and west, all came to the same conclusion that it must be (!) in Skaptárjökull... and according to the different points of observation, and the statement of our newspaper (!) at Reykjavik, the position of the crater ought to be between 64° 7' and 64° 18' north lat."'

The latitude of *Reykjavík* is 64° 9', therefore a

'crater' in 64° 7' would bear very slightly *south* of east (true), and if in 64° 18', of course but a trifle north of east, while if it bore east-north-east (22° 30' north of east) as stated, it would be within 1° 30' of the bearing of *Askja* from *Reykjavik*. Therefore, there cannot exist a doubt that the eruption seen in January from the capital was the one following the explosion within the chief of the *Dýngjufjöll* on the 4th of that month causing the earthquake.

It appears Burton did learn 'subsequently' more about the eruption, for he saw Mr. Watts upon his return from *Askja*, and upon the map attached to his book he shows in red ink the site of the *Askja* eruption on the western outer (!) side of the *Dýngjufjöll*, and the eruption on the 29th March, the great pumice eruption, as having taken place in the *Mývatn's Öræfi!!*

The letter bears date the 23rd March, and was, therefore, written five days prior to the great pumice eruption. From its tenor it is evident that had not the pumice eruption on the 29th March taken place, or Jón of *Víðikær* visited *Askja*, the site of the volcano active would never have been known, and the *Skaptár Jökull* would have been credited with the eruptions earlier in the year.

It is hardly to be wondered at, therefore, that not only is *Hekla*, the *Vatna Jökull*, and other volcanic vents credited with eruptions which have taken place elsewhere, but also that in the annals of Iceland we find no mention of any eruption among the *Dýngjufjöll*. Eruptions that took place from the volcanic mountains in the *Ódáðahraun* were, when recorded, mostly placed to the credit of the *Trölladýngja*, frequently under the name of *Skjaldbreið*; accordingly we find it stated that this volcano erupted in the years A.D. 1151, 1188, 1340, 1360, 1389, and 1510. In 1341 and 1510 we also find it mentioned that *Herðubreið*, a mountain twelve miles north-east of *Askja*, erupted; but Professor Johnstrup, who has examined *Herðubreið*, though I believe not made the ascent of the mountain, says this cannot be true as it is not volcanic. In 1477 and 1598 great eruptions of sand and pumice took place from some volcano or volcanoes in the interior, but which is not recorded. These were followed in 1618, 1862, and 1872 by violent earthquakes in the north; and in 1638, 1744, and 1862 we find it on record that flames were seen ascending in the desert.

It is highly probable that some, perhaps most, of these eruptions took place in *Askja*, as it is certain from the present condition of this vast crater, and the outer slopes of the mountain in which it lies, that they have been the seat of eruption after eruption during historical times. The flames that were seen ascending from the desert doubtless marked outbursts of lava from channels radiating from the volcanic vent below *Askja* underlying the *Ódáðahraun*, similar to the one that burst forth in the *Mývatn's Öræfi* in 1875. That such did issue in about the years mentioned, the newer lava-beds seen in crossing the desert conclusively prove.

The 1875 eruptions by exposing the widely-extending sheets of basaltic lava lying beneath the *Mývatn's*

THE PROBABLE GENESIS OF ASKJA AND ICELAND. 81

Öræfi, and making known the existence of a vast volcanic vent amidst the unexplored lava desert in the interior, to which we are enabled to trace the lava-flood that welled forth from a rift in the midst of a plain many miles distant from the volcano, afford in my humble opinion a clue to the post-tertiary geological history of the island; and, although no professor of geology, I will venture to translate 'the record of the rocks,' and repeat the story that the physical features of Iceland impart to me.

It was for a long time assumed that a fissure filled in with trachytic-lava bisected the island from S.W. to N.E., and that all the volcanic vents of Iceland lay upon this fissure; but as the interior became better known to modern scientists this theory was exploded, it being impossible to travel any distance without falling in with beds of lava that had welled forth through rifts in the more superficial rocky strata, presumably above channels in the substrata; while, moreover, in every part of the island volcanic mountains were met with that had erupted in comparatively recent times, and, as we shall presently see, there is a chain of volcanoes stretching across the island from south to north. Still, no great central vent was suspected in the interior of the island, although in the last few years a vague belief has prevailed that the various volcanic mountains that had been active since the settlement of the island were connected in some manner. That a central vent with radiating channels does exist is, however, now quite certain; and it is moreover equally so that the

G

vast sheets of basaltic lava which we saw bared in the *Öræfi*, and which underlie the *Ódáðahraun*, issued from this vent. It will be remembered that on the way to *Askja* we saw what the last that spread around must have been like, by the manner in which its surface had cooled; a veritable ocean of molten rock, almost or quite stationary, which had congealed evenly save where gigantic dome-shaped bubbles were formed by gases generated within or below the heated mass.

The presence of these vast lava deposits is conclusive evidence that the portion of the interior mentioned had, at one time, far less elevation than at present, and that it has been builded up by the deposition of the sheets of rock much in the same manner as a huge cauldron-shaped hollow in the heart of the *Djúpajufjäll* in more recent times has been filled up with smaller ones. I believe, indeed, that the greater part of the island of a less elevation than 1,500 feet has been builded up in this manner, as everywhere from sea level to this altitude, save where tracts of detritus brought down by the glacier and surface water are met with near the coast, the formation where exposed by earthquake rifts and river beds appears to be the same.

As observed in the second chapter, extensive tracts of the island, as well in the coastal regions as inland, have a far greater altitude, and are unquestionably an older formation than the plateau of the interior, and other less elevated parts of the island. The coastal region on the east and north mainly consists of semi-detached flat-topped mountain masses varying little in altitude, about 2,000 feet, with here and there tracts far more mountainous and much broken up, where similar masses have been tilted aside, the upheaved portions raised to a greater altitude by several hundred feet than the summits of their fellows, as also is the case where other masses have been bodily uplifted, while in some places masses lie lower, having apparently subsided. A mass of this older formation that has been tilted inland, is to be seen directly opposite *Akureyri*, with the angle at which it lies clearly shown by the weathering of the softer strata.

The whole of the north-west peninsula is a mass of this older formation, and it is believed *that nowhere upon it has a lava-flood welled forth in post-tertiary times;* a most important fact to be borne in mind when considering the geological history of Iceland.

In the interior and south part of the island there are also extensive tracts of this older formation, but they have a greater altitude than in the north, and are mostly covered with glaciers; therefore a glance at Gunnlaugsson's large map, where the *Jöklar* are coloured blue, will show in an instant their position. There are, however, in every part of the island, save the north-west peninsula and other tracts of the older plateau, smaller masses that are not marked as *Jöklar*, and some volcanic mountains of post-tertiary formation that are. In the west it is not an easy matter to point out upon the map the older from the newer formation farther seaward in that direction than the *Eyriks Jökull, Láng Jökull, Hloðufell, Torfa*

Jökull, and those to the southward, all, or most of which are masses of the older formation.

The flat-topped mountain masses of the coast, and the larger ice-clad *Jöklar* of the interior, are portions, in all probability, of one of the basaltic plateaux of north-western Europe (respecting which Professor Geikie contributed an excellent paper to a recent number of 'Nature'), a miocene, or middle tertiary, formation, while the plateau in the interior of the island, to the depth at least of the strata bared by rifts, is post tertiary beyond a doubt.

Throughout this little work I have endeavoured as much as possible to avoid making use of geological technicalities likely to mystify the general reader. I find, however, in this concluding chapter, that it is impossible to describe the probable genesis of Iceland without making use of a few. Therefore, I think it will be well here, in the interest of non-geological readers, to explain that throughout this chapter tertiary is the term applied to the igneous rock formations of Iceland that had their existence prior to the glacial epoch—about which I will say a few words presently, and post-tertiary to all formations of a later date.

Geologists assign many changes of the earth's surface, a number of rock formations, and a long duration to what they term the tertiary period, and bring it to a close with the glacial epoch. This epoch was one of great disturbance of the earth's crust from other causes than the action of the ice, subsidences and upheavals to the extent of 1,800 to 2,000 feet taking place over the whole of the northern hemisphere from the 40th or 42nd parallel northward.

Prior to this epoch, there is every reason to believe, there existed north of Scotland an extensive plateau, extending far into the Arctic sea, built up of igneous rocks—tuffs, basaltic and other lavas—which had welled forth from rifts in the earlier rock formations of the earth's crust, and been deposited in horizontal strata one above the other in a precisely similar manner to the sheets of basaltic-lava underlying the *Mývatn's Öræfi*. Professor Geikie is of opinion that the hills of Antrim, Mull, Morven, and Skye, the Farœs, and part of Iceland are surviving fragments of this formation; and there can exist no reasonable doubt that he is right, while to judge from the photographs of the cliffs of some islands in the Arctic seas, taken during Mr. Leigh Smith's very successful cruise last summer, there are fragments still further north.

There were, doubtless, numerous vents in the earth's crust through which these vast sheets of molten rock welled forth, as similar formations are found on the western continent in the same latitudes, and elsewhere; but whether they were longitudinal, or latitudinal rifts of any great length, there is no evidence to show. However, one thing is tolerably certain—Iceland lies above one of these vents and marks a spot where, subsequently to the glacial-epoch, igneous rock has welled forth and builded up a plateau very similar to the older one, though comparatively inconsiderable as to extent.

There is no dearth of evidence that the flat-topped

mountain masses on the north and east coasts are of far older formation than the inland plateau, nor that the low-lying portions of the coast and many of the higher volcanic mountains, and several of the ice-clad heights are coeval in formation with the interior. I have in various places, from Seyðisfjörðr on the east to the south of the north-west peninsula, examined the upper strata of the flat-topped mountain masses on the north coast, and *they do not correspond* with those of the existing inland plateau, the former having conglomerate and tuff strata alternating with basaltic and other lavas (in places three or four lava-strata will be found resting one upon the other with a stratum of tuff or conglomerate below and another above), whereas the uppermost strata of the latter, bared by the rifts to a depth of quite two hundred feet, are exclusively of lava that has been deposited unchanged, save by congelation, with thin layers of clinker-like fragments that have never been abraded by ice or waterworn, marking the divisions between; therefore it is evident these mountain masses, even if they were formerly a continuation of the substrata underlying the existing inland plateau, must have been upheaved prior to the deposition of the later lava-flows bared by the rifts, or the interior now covered by them must have subsided. Moreover, while the coastal valleys and fjord inlets show signs of glacial action, the surface of the superficial sheet of lava on the plateau has never been abraded by ice, otherwise the rounded summits of the dome-shaped bubbles seen in the Ódáðahraun and elsewhere, would have been ground away: therefore it is certain the later ones must have issued subsequently to the glacial epoch: that they issued subærially and did not course down an inclined plane, but spread out in an immense basin inclosed—or more strictly, as many gaps doubtless existed, partly inclosed—by the fragments of the miocene plateau is likewise certain, or they would not have been deposited so horizontally and evenly.

Further proof is to be found in the facts that deep narrow fjords exist only on the east, north, and northwest coasts, where this older formation prevails, and not on the south and west, where the island is built up to a greater extent by the igneous rock which has coursed between the masses of the miocene formation —the ice-clad *Jöklar* lying inland to the south and west of the elevated plateau of the *Sprengisandr* and *Ódáðahraun*. The way in which the fjord inlets radiate is convincing proof that they have been mostly hollowed out by glacial action; and in the north-west peninsula the glaciers are still doing their work, though perhaps not so vigorously as of yore, the glaciers being of less extent. Save on the north coast —where we find four, whose beds and continuous valleys lie between detached masses of the older formation, the fjords nowhere have *deep* continuous valleys that penetrate inland beyond the more elevated coastal region; and the river valleys, even those of the *Skjalfandafljót* and *Jökulsá*, rivers which intersect the island northward from the very base of the *Vatna Jökull*, are of no great depth upon the plateau, but abruptly deepen as the coast is approached, the two rivers named each falling nearly a thousand feet in a few miles.

While upon the subject of the strata underlying the plateau bared in these long waterworn river beds, I will digress to briefly call attention to the way in which the geological student at the very commencement of his career is puzzled by the present classification of rocks,* by asking: In the face of the vast stratified formations of igneous rocks that have issued, and been deposited, subærially, found all over the globe, is it not about time that such an absurd dogma as the one embraced in the following paragraphs was erased from works on geology?

'There are thus in the crust of the globe only two great categories of rocks—the aqueous or stratified, and the igneous or unstratified; the former produced through and by the agency of water, the latter through and by the agency of fire.' 'Here, then, as we cannot regard nature acting in time past otherwise than at present, we are entitled to infer that all (!) rocks in the earth's crust occurring in layers have been formed through and by the agency of water.' The above are from 'Geology for General

---

* The term 'trap rock' is also very confusing. Dana applies the name to a 'dark greenish or brownish-black rock, heavy and tough. Specific gravity 2·8—3·2,' while Page says, 'Trap-rock (so called from the step-like aspect it gives to hills composed of it) is a name which includes a great variety of igneous rocks, the general characteristics of which are easily (?) recognised in the field. Basalt and greenstone may be included under the term trap, but the name is generally applied to the looser and less crystallised masses, known as trap-tuff, wackè, amygdaloid, &c.'

Readers,' and I can only observe that it is a matter for regret that the author did not devote less time to theoretical indoor study of the science of geology, and more to practical outdoor study in such lands as Iceland. The geological dogma that all stratified rocks are of subaqueous deposition has proved an incubus that previous writers when describing the geology of Iceland have not been able to shake off, not even Burton, for he guards his statements by the following footnote:—' The word " trap " will be used in these pages to denote the lavas ejected by submarine volcanoes.'

To judge by the existing coastal formation of Iceland, the masses of the miocene plateau left standing subsequently to the disturbances of the glacial epoch were detached by arms of the sea as the Faroes are to day, or what is perhaps more likely were separated by the arms of a vast glacier occupying the space now the interior of the island, which stretching seaward through rifts in and the gaps between the mountain masses, deepened and widened them into the fjord inlets of this remarkably indented island. Amidst the complex of mountainous isles left standing, existed a vast volcanic outlet that has remained active until this very hour; belching forth its fiery floods of molten rock to be deposited in the form of conglomerates and tuffs when they issued subaqueously or subglacially, and later on, subærially, in the form of the vast sheets of lava composing the more superficial strata of the inland plateau. Not only in the large space inclosed by the fragments of the miocene plateau was the molten rock so deposited, but coursing between the rifts and gaps a certain quantity was left behind, the sea and ice thereby being gradually ousted, and the detached masses connected by lower lying tracts into one island, with an exceedingly irregular coastal outline, and thus was the foundation laid of the Iceland of to-day.

There is reason to believe that the sea-level over the northern hemisphere varied considerably towards the close of the glacial epoch, and likewise during the early ages of the post-tertiary period, at times being far higher than at others. This would account for the tuff-strata found alternating with the lava; as it is not very speculative to imagine that those portions of Iceland, including the interior, lowlying at that time would be flooded; the volcanic vent converted for a period into a submarine volcano, and the molten rock that issued at such times be deposited in the form of tuffs and conglomerates. In the course of ages the whole of the spaces between the fragments of the miocene plateau were gradually upbuilded above the encroachment of the sea; the currents and tides carrying away the greater part of the disintegrated igneous rock that found its way beyond the outlying miocene bulwarks that now form a considerable part of the coastal region; the climate became more temperate, no glaciers being found at a lesser altitude than 2,500 feet, and the later discharges of molten rock were deposited unchanged, save by congelation, in the form of the basaltic and other lava-strata that lie uppermost in the post-tertiary plateau forming the greater part of the interior.

The regularity and evenness with which these vast sheets have been deposited seem to point to a remarkably peaceful welling forth of the floods of molten rock, accompanied by the discharge of little or no fragmentary material, for no layers of ash interpose between the strata, the thin layers of clinker-like crust alone being found. Their deposit, however, appears to have first greatly narrowed in, and ultimately to have sealed the vent whence they issued; and then came troublous times.

As to the site of this volcanic vent, I believe that immediately following the disturbances of the glacial epoch a rift in the earth's crust extended from south to north under those portions of the island now known as the *Vatna Jökull* and the *Ódáðahraun* as far north as *Krafla*; fragments of the miocene plateau being left standing on both sides of the southern end of the rift, it was narrower than in the centre, and the superincumbent masses of igneous rock now existing there, were sooner piled up than where the rift was wider; as also was the case towards the northern end where it also was less in width, the volcanic vent being thus at a comparatively early period of its post-tertiary history narrowed down to the limits of the *Ódáðahraun*. The mountain masses lying upon these rifts on the north and south, stand to this day as a glance at the map will show. The greater number of the foundation strata of the island were probably deposited during the time the outlet extended north and south through the Ódáðahraun, and by the welling forth of the later sheets of molten rock, the outlet was still further narrowed down to the limits of the *Askja* crater. The Ódáðahraun lies higher than the *Mývatn's Öræfi*, strong presumptive evidence that it lies nearer the vent whence the sheets of rock issued, and also that the later ones were less in bulk than those preceding, and consequently did not extend farther round the vent than the borders of the Ódáðahraun; the very last that issued prior to the upbuilding of *Askja's* mountainous periphery being the comparatively smaller ones, that form its widely extending base. One thing is certain, *Askja* is surrounded with a tract covered with lava which, subsequently to the deposit of the most superficial sheet of rock on the plateau, has issued from that crater, and welled forth from rifts and smaller volcanic mountains around it, having as large an area as the whole of the other tracts similarly overspread in the island; and this fact in itself is strongly corroborative of the view taken that the *Askja* crater marks the focus of volcanic activity in Iceland.

It is an easy matter to trace the rift, of which *Askja* is the present great central outlet, for it forms to this very hour a covered channel running almost due north for a distance of forty miles, and another running south to the coast, both clearly traceable by signs of active volcanicity in several places, where gases from the central vent force their way through the porous lava filling in the rift, and form solfatarar in spots where cracks and fissures exist.

The *Fremrinámar* (Farther-solfatarar) lie upon the northern channel on the very verge of the *Ódáðahraun*, and mark where an immense flood of lava in comparatively recent times burst forth through the overlying strata, roofing in the rift, at a weak spot between the site of former disturbance farther north, where volcanic mountains and vast lava-floods piled above offered more resistance, and another volcanic range, the Northern *Dýngjufjöll*, which mark the course of the rift through the *Ódáðahraun*. Farther north, upon the line of this rift are also the *Illiðarnámar*, the solfatarar in which are situated the mud wells briefly described in Chapter II, *Leirhnúkr*, and *Krafla*; and it is possible at the latter place, which has been the scene of terribly violent eruptions, ramifications extend to the *þeistareykir* solfatara and the *Uxahver* group of hot springs. From *Leirhnúkr*, and a number of fissures between that volcanic mountain and the *Fremrinámar* (all lying in a line running from south to north upon the rift), lava-flood after lava-flood streamed forth in the years A.D. 1724, 1725, 1727, 1728 and 1729, spreading over the plateau in both an easterly and westerly direction. That portion of the rift south of *Askja* runs nearly due south under the *Ódáðahraun*, and the icy wastes of the *Vatna Jökull* to the coast, its course being marked by the *Kverkhnákarani* and the *Kverkfjöll*, a range of volcanic mountains penetrating far amid the glaciers of the *Jökull*, possibly even as far south as the *Œræfa Jökull*, the volcano which marks the termination of the rift. From the *Kverkfjöll*, Mr.

Watts saw smoke ascending during the *Askja* eruption in 1875. Further, and most convincing evidence that the volcanic vents on this rift are connected with a central vent, and with each other, is found in the phenomena of the eruptions in the years 1727, 1728, and 1729. It is recorded that the *Oræfa Jökull* took the initiate on the 3rd August, 1727, *Leirhnúkr* following suit on the 21st of that month, the *Jökull* remaining active until the spring of the following year, when a fresh outburst at *Leirhnúkr* on the 18th April, and the opening of rifts and the formation of erupting craters in the *Dalfjall*, in *Hrossadalr*, and at *Bjarnarflag*—all three spots close together on the rift between *Leirhnúkr* and *Askja*, and lying at a far lower altitude than the *Oræfa Jökull*—the lava-floods were diverted from the *Jökull* and found outlets in the four places mentioned. *Leirhnúkr* was active during the whole of the following year. It is also worthy of notice, that in 1728, when no less than five volcanic vents were active upon a clearly traceable rift, or channel, running north and south from *Askja*, it is recorded that 'this same year volcanic action was going on in the lava wastes round *Hekla*.' It will be remembered that I mentioned on page 15, that to judge by the 'lay' of most of the lava-beds around that volcano they appeared to have welled forth from fissures above a channel running from S.W. to N.E. that would intersect the *Askja* crater, if such existed and extended far enough in a north-easterly direction, and not from *Hekla* or from rifts radiating from the volcanic vent of that mountain; and the fact that 'volcanic action was going on in the lava wastes round *Hekla*,' and not from the volcano itself, at the time that five vents upon a far distant rift or channel were erupting is, at any rate, something more than slightly corroborative of the opinion expressed by me that a channel also runs in the direction of *Hekla* from a main central vent.

From the number of lava-floods that have burst forth at various times at a distance from volcanic mountains in the same way as the one in 1875, it is evident that innumerable channels exist in the post-tertiary strata; and, to judge from the phenomena of the eruptions in the three years from 1727 to 1729, and those in 1875, there is every reason to believe that these channels are connected with the great central volcanic vent beneath *Askja*.

A remarkable proof of the existence of radiating channels is found in the fact, that during the eruption in the vicinity of the *Skaptár Jökull*, in 1783, of the most prodigious lava-flood of which we have any record in Iceland or elsewhere, lava welled forth beneath the sea and built up an islet, *Eldey*, a few miles south-west of *Reykjanes*, over one hundred and fifty miles from the *Skaptár Jökull*!

To show that I am not singular in my belief that the volcanic vents of Iceland are connected, I will quote a paragraph from Mr. Watt's work:—

'In journeying through these centres of volcanic activity, we cannot but be struck with the general lowness of the volcanoes in Iceland. This is doubtless owing to the number of vents which exist in close proximity to one another, so that the volcanic force, having piled up a certain amount of superincumbent matter, finds readier exit by bursting through the superficial overlying rocks in adjacent localities, which offered less resistance than the accumulated volcanic products which they themselves had previously erupted, or by availing themselves of some pre-existing point of disturbance which afforded them a readier escape.'

It is not a very difficult matter to propound a feasible geological theorem to account for the existence of subterraneous channels in an island built up, as Iceland has been, chiefly by volcanic agency. The earlier discharges of molten rock from the great central vent would in many places find outlets to the sea between the detached outlying masses of the miocene plateau, and, the climate being hyperborean, thick crusts would form upon the fiery floods even as they flowed and create covered channels, through which the molten rock would continue to course; and being protected from the cold, it would remain in such an exceedingly molten state that nearly the whole of it would drain out, leaving the channels empty, save that each would be blocked in places, more especially near its outlet, by slag and clinkers, the dregs of the molten streams, and possibly also by subsidences of portions of the roof. The great central vent not being, as yet, greatly narrowed in by the deposit of vast sheets of rock around it, the next molten flood that issued would well forth quietly, but an inconsiderable quantity force its way past the slag and clinkers into the channels, but what did would be there congealed and further seal them; and this sheet of molten rock would be followed by a succession of others, each issuing at a greater altitude than the one preceding, and some would be certain to drain out through higher lying gaps in the miocene mountain masses, and form similar but smaller channels radiating from the central vent in a different direction to those formed earlier in the history of the island; as well as in some instances above existing channels: moreover some of the latter would be enlarged by the molten rock coursing over and redissolving their rocky roofs, and in most instances these enlarged channels would likewise be roofed in by the congelation of the surface of the later molten floods. This probably went on for ages; the deposition of these vast sheets of rock by degrees narrowing the dimensions of the vent, while the molten floods that issued became less and less in bulk, so that each succeeding one did not flow as far as the one preceding, and being deposited in strata above the earlier deposits the dimensions of the central vent were still further reduced, and it was finally closely sealed by a rocky core formed of the congealed dregs of the last molten flood that welled forth peacefully. These last deposits are the more superficial of the sheets of lava underlying the *Ódáðahraun*, the surface one being, as before said, about 200 feet higher than that in the *Mívatn's Öræfi*.

The period following the close sealing of the great central vent was, beyond doubt, the most troublous one in the history of the island since the glacial epoch; most of the mountains, not portions of the miocene plateau, now standing in the island being formed by the volcanic disturbances that succeeded. Let us try to picture in our minds what occurred upon the very next occasion that a flood of molten rock sought to force an outlet. The pressure of the confined gases generated from millions of cubic feet of molten matter is great within the sealed vent; the masses of slag and clinkers blocking the channels radiating therefrom are more easily forced aside than the rocky core sealing the outlet, and the molten matter forces its way into and courses through one or more of the channels. On rush these floods of fluid fire, till, at spots not far distant from the sea where the channels are more closely sealed by the congelation of the tail ends of former molten floods, water is met with that has percolated through the porous igneous rock and lodged in the channels, an immense quantity of steam is suddenly generated at each spot and an explosion ensues, the rocky core between the molten matter and the outlet, and the continuous molten flood behind in the channel, with the vast deposits lying above increasing in thickness as the sealed central vent is neared, offer more resistance than the less numerous strata immediately above where the explosion takes place, and here a mass of the rocky strata is tilted into the sea, or upheaved in fragments and a mountain built up, to be for a time an active volcano, and in after years, when the heated volcanic matters are diverted from, or have sealed this particular channel, an ice-clad *Jökull*. And thus, I believe, was the coastal region of post-tertiary formation, especially on the west and southwest, much broken up, and the volcanic mountains now standing on the lowlying land near the sea, upbuilded subsequently to the formation of the plateau in the interior.

Reykjavík stands upon a tract that has been torn away from the main body of the island in comparatively recent times, which is even now slowly sinking. There are old men still living whose fathers were born in houses of which, at the present time, the foundations only are visible at low water on a spit of land jutting into the bay west of the town. Tourists who have 'done' Þingvellir and the *Geysir* will remember that shortly after fording a river a few miles from *Reykjavík* their route lay among a number of hills, and that they ascended by steep and crooked ruts—paths would be a misnomer, sometimes the beds of mountain streams, to the *Mosfellsheiði*. These hills are above the line of disruption, and the large floods of lava seen between the capital and *Hafnarfjörðr*, doubtless issued from rifts at the time the tract was torn away.

The majority of the volcanic mountains inland each mark a weak spot above a radiating channel, or where one was blocked by a core of congealed lava, and were formed in the same way as those near the coast, surface water from the snows and rains finding its way into the channels in place of water from the sea.

I believe that the molten rock from the central vent, even when not forced into the channels by the pressure of confined gases, slowly 'eats' its way by redissolving the igneous rock lying in them, and that when it lies any length of time in a large channel this is enlarged by the melting of its walls and roof, which yield as readily to the intense heat of a large quantity of molten rock as a rod of soder would if inserted into a crucible of molten lead: therefore at this very hour it is likely that in many of the channels radiating from the vent of which *Askja* is the outlet, molten matter is slowly and steadily redissolving the lava with which they are clogged and inclosed. It is fearful to think that there are 70,000 people living upon an island honeycombed with channels charged with molten rock that may at any moment burst forth beneath a homestead, and consume it and its inhabitants. Equally alarming is the fact that in the crater above the great central volcanic vent is a large lake of recent formation, whose water—year after year increasing in volume, and now five miles in circumference and two hundred feet in depth—will assuredly, should a slight explosion loosen the lava-floor of that crater, find its way below, come in contact with molten matter, and cause an explosion to which those in 1872 and 1875, which opened rifts from ten to thirty miles in length, will be comparatively insignificant, an explosion that in all probability will send another molten flood coursing over the island—and who can tell where? it may be in the very streets of *Reykjavik* itself!

The history of *Askja*, subsequently to the building up of the inland plateau and the uniting thereby of the miocene fragments into one island, may be told in a few lines. Let us imagine that the deposits immediately around the now greatly narrowed in and sealed central vent have been several times ruptured by explosions accompanying the eruptions which built up the volcanic mountains now standing in every part of Iceland of post-tertiary formation; and that the lava-flows now forming the widely extending base of *Askja's* encircling mountain wall have issued from the fissures around the immense core of congealed lava plugging the central vent, whose diameter is that of the present *Askja* crater; that the radiating channels are mostly hermetically sealed by the congealed remains of former molten floods, and that the masses of tuff and other strata now forming the highest part of *Askja's* encircling mountain have yet to be upheaved. A tide of molten rock again sets in the direction of the Icelandic outlet; and ere it can 'eat' its way into the radiating channels, the disturbed strata surrounding the core are partly redissolved by the intense heat, fissured asunder, and torn therefrom by the enormous pressure below, and the core, being no longer supported by their cohesion, sinks bodily to some depth into the abyss, in much the same manner as the tract now beneath the lake did in 1875, and the deep cauldron-shaped hollow was formed in which the lava-floods bared in the cliffs bordering the lake on the north were to be subsequently deposited. The eminence formed by the summit of the core and the surrounding strata was doubtless, owing to its altitude, covered with snow, and this melting, the water came into contact with the molten matter, adding by the generation of an immense volume of steam to the violence of the terrible convulsion that took place when the huge masses of the tuff sub-strata lying far below were upheaved through the innumerable sheets of lava lying above them, and *Askja's* mountainous periphery was built up above the base formed of these deposits, by the masses of strata upheaved, and the amorphous matter which issued during the eruption. Possibly, all around the core as it subsided, lava welled up and spread over its summit within the cauldron-shaped hollow, the first stratum to be deposited therein. In course of time it is likely the core would be gradually redissolved by the intense heat of the molten matter by which it was surrounded, and ejected in the form of lava, as only by such action can the existence be accounted for of the abyss into which the mass of strata, five miles in circumference, disrupted in 1875 subsided. Lava-flood after lava-flood has subsequently burst forth in the crater, and in the course of ages it has been filled up to its present level, that of the two gaps in its mountainous periphery, through the eastern of which, at any rate, the later lava-floods that have here issued have found an outlet and flowed over the plain around.

This, I respectfully submit, is the probable history of the volcano, which in the preceding pages I have attempted to describe.

Hitherto, with the exception of the *Ódáðahraun* and the *Vatna Jökull*, I have said but little with reference to the extent of the lava-covered tracts and immense glaciers of Iceland, nor mentioned the area of the deserts of volcanic sand and rocky débris lying in the interior west of the *Ódáðahraun*, or of the elevated moorlands where the Icelanders pasture their sheep in summer. I will now give these particulars, as their omission would render this brief and incomplete geological account of the island still more so. The area of Iceland is given by Gunnlaugsson as 1,867 Danish square miles = about 38,000 English: nearly one-seventh of which consists of *Jöklar*, or glacier-covered heights, the *Vatna Jökull* being credited, as before stated, with an area of 3,000 square miles, the *Hofs Jökull* with 500, *Láng Jökull* with 440, the group of *Jöklar* in the south, of which the *Mýrdal's Jökull* is the chief, with about 350, and the *Dránga* and *Glámu Jöklar*, lying on the north-west peninsula, with 400. The ice-clad mountain masses mentioned will be found upon examination, I think, to consist mainly of fragments of the miocene plateau, though in the *Vatna Jökull* group, it is certain, there are some of post-tertiary formation, but these have been before alluded to. The area of the lava deserts, *i.e.* tracts covered with rugged beds of lava that has welled forth from rifts or flowed from volcanic mountains in quite recent times (hundreds of miles have been so covered since the settlement of Iceland), is computed at 2,400 square miles, of which the *Ódáðahraun* is credited with half, while the lava-covered tract round *Hekla* is said to be but 240 square miles in extent. In the south-western portion of the island, between the capital and the south coast, there are over 500 square miles of country covered with lava; and between the *Mýrdal's Jökull* and the *Skaptár Jökull* there are the tracts covered with the lava-floods that there streamed forth in 1783, two entire river valleys, and over 200 square miles of lowlying land!

The area of the sandy, stony deserts of the interior may be set down at the least at 5,000 square miles, and they are slowly but surely increasing in extent, as will be hereafter seen. The two largest are the *Sprengisandr* and *Stórisandr*, and these cover the surface of the elevated inland plateau on the west of the *Ódáðahraun*, from which the one first named is separated by the *Skjalfandafljót*. The boundaries of the two deserts where they join are the rivers draining the *Láng* and *Hofs Jöklar*, whose miocene foundations rise through the post-tertiary plateau on the south of the deserts: in each of which, by-the-bye, is a tract about thirty square miles in extent, covered with lava that has welled forth from rifts. I must not omit to state that south of the *Vatna Jökull* there is a tract of sand quite 200 square miles in extent, lying but little above sea-level; it is named the *Skeiðarársandr*, and is the detritus of the glaciers.

With reference to the moorlands and pastures (there is no arable land), Professor Johnstrup says: 'It is exceedingly difficult to give, even approximately, an estimate of the area of the pasture lands of Iceland, as these imperceptibly run into extensive heaths with a most meagre grass growth. The area of the pasture land is usually estimated at 7·46 (Danish =15,000 English) square miles, but this includes all that can by any means be brought under this heading.'

This is, unquestionably, an over-estimate, notwithstanding that a large extent—possibly 3 to 4,000 square miles—of the eastern part of the island is elevated moorland, for year after year the area of the pasture land is decreasing, while that of the sand deserts is increasing—every gale of wind sweeping over the latter, when uncovered by snow, spreading over the former a thin covering of wind-borne sand; and it is impossible to travel far, especially in the vicinity of the larger deserts, without coming across extensive tracts of moorland buried to a depth of several inches, which but a few years since nourished a growth of stunted birch and willows, whose stems project through the black sand bleached and dead. With reference to the thousand square miles or so of moorland in the east, buried under pumice by the 1875 eruptions, I think they will only remain useless as pasture land for a few years, the expansive action of the frost upon the rain and snow water within the porous pumice will soon degrade it into fertilizing soil, and the grass growth here will be richer than ever. There is no great depth of soil upon the elevated moorlands: we saw in the *Mývatn's Öræfi* that there was but a depth of five feet, and that is about the average.

The glaciers, lava and sand deserts, moorlands, and pastures thus cover about 28,000 square miles, and the remaining 10,000 consist chiefly of mountain masses varying in altitude from 2,000 to 3,500 feet, snow-covered for nine months out of the twelve.

From the above account of the superficial physical

features of the island, and the description of the journey inland to *Askja*, the reader will be able to form some idea of Iceland; of the wild weirdness of its landscapes, where widely extending ice-clad mountains rise from fire-blasted deserts of lava, volcanic-ash and sand, resting upon an elevated plateau whose rocky foundation is shivered in all directions with deep earthquake rifts; of the ice and snow-clad remnants of the miocene plateau, whose rocky buttresses on the north have withstood the raging of the Arctic waves through sunlit summer night and sunless winter day from a period when man was not; of the grassy oases amid the black and sterile plains where stand the homesteads of the Icelanders; a race as patriotic as the Swiss, and more hardy, who, in defiance of volcanic eruptions and earthquakes, and a hyperborean climate, cling to the island home of their forefathers, and live a peaceful, contented, pastoral life; worshipping God with all the earnest simplicity of the Lutheran religion, and hospitably entertaining the stranger that may come among them, placing at his disposal the best room in the house, which is everywhere reserved as a 'guest-room.'

London: J. S. Levin, Steam Printing Works, 2, Mark Lane Square, E.C.

WORKS BY THE AUTHOR.

NORWAY.
THE SPORTSMAN'S HANDBOOK.

Crown 8vo., 376 Pages, cloth, bevelled boards, 10s. 6d.; by post 11s.

'SPORTING LIFE ON THE NORWEGIAN FJELDS'
('TILFJELDS').

By Prof. FRIIS, with a MAP printed in colours, showing where ELK, REDDEER, and REINDEER are to be found. An ENGLISH TRANSLATION of the above by W. G. LOCK, with a SYNOPSIS of the NORWEGIAN GAME LAWS, and chapters on the RENTAL of SHOOTINGS, SALMON FISHING, ETC., forming the most perfect SPORTSMAN'S GUIDE TO NORWAY yet published.

'THE FIELD,' in its Review of 'TILFJELDS,' as a Norwegian work, stated :—

'Regarded by his countrymen as one of the most accomplished and successful of the newer school of Norwegian sportsmen, Herr Friis volunteers to act as guide to districts, where his wanderings, extended over more than twenty summers, have rendered him familiar with every streamlet, tarn or waste every haunt of fish or game. He would appear, at some time or other, to have extended his search for game over all the principal fjelds in the south of Norway and to be nearly as much at home on Dovre, or on the Lesje Fjelds, in Jotunheim, or on the Hardanger-vidde. The traveller . . . will find Herr Friis's book . . . well worthy of a place among his baggage . . . as a guide book, or also as a pleasant resource in times of unavoidable idleness.'

'THE FIELD,' thus Reviewed the ENGLISH VERSION :—

'We congratulate the numerous class of Anglo-Scandinavian sportsmen who have not mastered Norsk on the very readable addition to their literature afforded by the above translation, for which we have to thank Mr. Lock. A review of Professor Friis's original work appeared in our number of 24th of March, 1877 (p. 347), and to the favourable opinion therein expressed we can only add that the English version is faithful and fluent. Professor Friis is such a master of his subject, and so enthusiastic and yet simple in his treatment of it, that Mr. Lock, who is also a sportsman and lover of the rifle, must indeed have found it a labour of love to Anglicise the book for the benefit of his fellow islanders. Indeed, the work can be safely recommended for perusal by the non-sporting public, as the purely shooting and fishing episodes are so pleasingly interwoven with descriptions of scenery, natural history, and botanical notes, and happy illustrations of the inner life of peasants, milkmaids, and other minor *dramatis personæ*, that there can be no lack of interest. There is, moreover, a vein of quiet humour in the author, which at times is quaintly comical. . . .

'Mr. Lock, besides some useful explanatory notes, adds hints to English sportsmen visiting Norway, accounts of rental of shootings, instructions for salmon fishing, and a synopsis of the Norwegian game laws, so often of late discussed in our columns.

'The map referred to in the title shows by coloured patches where reindeer, elk, and reddeer are found in Southern Norway—somewhat on the same scheme as the excellent maps of Schübeler and Collett, which respectively illustrate botanical and zoological distribution in the Scandinavian peninsula.'

[OVER.]

WORKS BY THE AUTHOR (continued).

'THE LIVE STOCK JOURNAL'S' critique was as follows:—

'Now that every one goes to Norway at some time or other, all books—and they are growing—on the sport and adventure to be had on its fjelds and fjords are eagerly sought after. The one before us now is a translation of Professor Friis's "Tilfjelds," a work of the highest authority, and one which every sportsman visiting Norway ought to have. In its English dress it will be available to a large number to whom in the original it has hitherto been a sealed book. Mr. Lock won the copy from which he made his translation as a prize in a sort of Arctic Wimbledon in which he took part against Norwegian riflemen at Harstad. Save for the honesty which confesses it, no one would guess this was a translation, it reads so fluently and easily.'

PUBLISHED BY THE TRANSLATOR, at 16, Kingston Terrace, Charlton, S.E. Only a few copies now Remaining on Hand.

ICELAND.

Crown 8vo. 184 pages, cloth, limp. Price 5s. net.

GUIDE TO ICELAND.

A USEFUL HANDBOOK for TRAVELLERS AND SPORTSMEN, with a LARGE MAP showing EVERY RECORDED SITE OF VOLCANIC ACTIVITY, PLACES OF INTEREST, SALMON RIVERS, REINDEER TRACTS, FARMS WHERE NIGHT QUARTERS ARE OBTAINABLE, ROUTES, etc., by W. G. LOCK, F.R.G.S.

CONTENTS.

CHAPTER I.—INTRODUCTION.

Section I.—General Information . . 1
  „ II.—Outfit, Guides & Ponies . 22

CHAPTER II.—SUCCINCT ACCOUNT OF ICELAND.

Section I.—Brief Topographical & Geological Notice . 37
  „ II.—The People and the Climate . . . 49
  „ III.—Historical Notice and Outline of 'The Njál Saga' . . 53

CHAPTER III.—NOTES ON SPORT.

Section I.—Shooting 75

Section II.—Angling, and List of Icelandic Salmon Rivers and Chief Trout Streams . . 80

CHAPTER IV.—THE CAPITAL AND VICINITY.

Section I.—The Capital . . . 87
  „ II.—Excursions in its Vicinity 90

ROUTES.

Route I.—To Þingvellir, Geysir, and Hekla . . . 93
  „ II.—Through 'The Njál Country' . . . 116
  „ III.—A Summer's Tour . 124
  „ IV.—Up the West Coast . 162
  „ V.—Seyðisfjörðr to Akureyri 172
  „ VI.—Akureyri to Reykjavik 173
Conclusion.—Desert Routes . . 174

PUBLISHED BY THE AUTHOR at 16, Kingston Terrace, Charlton, S.E., who will forward copies on receipt of crossed cheque or P.O.O., payable at Lower Charlton Post Office.

BOOKS
PUBLISHED
BY JAMES PARKER AND CO.
OXFORD; AND 6, SOUTHAMPTON-ST., STRAND, LONDON.

Theological, &c.

REV. CANON HOLE.
HINTS TO PREACHERS, with SERMONS and ADDRESSES. By S. REYNOLDS HOLE, Canon of Lincoln. Post 8vo., cloth, 6s.
Second Edition nearly ready.

REV. E. B. PUSEY, D.D.
WHAT IS OF FAITH AS TO EVERLASTING PUNISHMENT? In Reply to Dr. Farrar's Challenge in his "Eternal Hope," 1879. By the Rev. E. B. PUSEY, D.D., Regius Professor of Hebrew, Canon of Christ Church. Fourth Thousand. 8vo., cloth, 286 pp., 3s. 6d.

REV. E. F. WILLIS, M.A.
THE WORSHIP OF THE OLD COVENANT CONSIDERED MORE ESPECIALLY IN RELATION TO THAT OF THE NEW. By the Rev. E. F. WILLIS, M.A., Vice-Principal of Cuddesdon Theological College. Post 8vo., cloth, 5s.

ARCHDEACON DENISON.
NOTES OF MY LIFE, 1805—1878. By GEORGE ANTHONY DENISON, Vicar of East Brent, 1845; Archdeacon of Taunton, 1851. Third Edition, 8vo., cloth, price 12s.

HERBERT DE LOSINGA.
THE FOUNDER OF NORWICH CATHEDRAL. The LIFE, LETTERS, and SERMONS of BISHOP HERBERT DE LOSINGA (b. circ. A.D. 1050, d. 1119). Edited by E. M. GOULBURN, D.D., Dean of Norwich, and H. SYMONDS, M.A., Rector of Tivetshall. 2 vols. 8vo., cloth, 30s.

THE LATE CANON JENKINS, D.D.
PASSAGES IN CHURCH HISTORY. Selected from the MSS. of the late Rev. JOHN DAVID JENKINS, D.D., Vicar of Aberdare; Canon of the Cathedral of Natal. With a Brief Memoir of the Author, by T. J. DYKE. Edited by F. M. F. S. 2 vols., Crown 8vo., cloth, pp. xxiv—1,108, 15s.

REV. J. WORDSWORTH, M.A.
UNIVERSITY SERMONS ON GOSPEL SUBJECTS. By the Rev. JOHN WORDSWORTH, M.A. Fcap., cloth, 2s. 6d.

HENRY HARRIS.
DEATH AND RESURRECTION. With an Introduction on the Value of External Evidence. By HENRY HARRIS, B.D. Fcap. 8vo., cloth, 3s. 6d.

BISHOP OF BARBADOS.
SERMONS PREACHED ON SPECIAL OCCASIONS. By JOHN MITCHINSON, Bishop of Barbados. Crown 8vo., cloth, 5s.

THE LATE BISHOP WILBERFORCE.
SERMONS PREACHED ON VARIOUS OCCASIONS. With a Preface by the Lord Bishop of ELY. 8vo., cloth, 7s. 6d.

REV. JOSEPH DODD, M.A.
CONSECRATION; or, A PLEA FOR THE DEAD. By the Rev. JOSEPH DODD, M.A. With an Appendix stating the Law with respect to Churchyards and Burial-grounds. By J. THEODORE DODD, M.A. 8vo., sewed, 165 pp., 4s.; in cloth, 5s.

THEOLOGICAL WORKS, &c. (continued).

THE BOOK OF COMMON PRAYER.

AN INTRODUCTION TO THE HISTORY OF THE SUCCESSIVE REVISIONS OF THE BOOK OF COMMON PRAYER. By JAMES PARKER, Hon. M.A. Oxon. Crown 8vo., cloth, 12s.

THE FIRST PRAYER-BOOK OF EDWARD VI., compared with the Successive Revisions of the Book of Common Prayer; with a Concordance and Index to the Rubrics in the several editions. By the same Author. Cr. 8vo., cl., 12s.

REV. E. B. PUSEY, D.D.

DANIEL THE PROPHET. Nine Lectures delivered in the Divinity School of the University of Oxford. With a new Preface. By E. B. PUSEY, D.D., &c. *Seventh Thousand.* 8vo., cloth, 10s. 6d.

THE MINOR PROPHETS; with a Commentary Explanatory and Practical, and Introductions to the Several Books. 4to., cloth, 31s. 6d.

THE DOCTRINE OF THE REAL PRESENCE, as contained in the Fathers from the death of St. John the Evangelist to the 4th General Council. By the Rev. E. B. PUSEY, D.D. 8vo., cloth, 7s. 6d.

THE REAL PRESENCE, the Doctrine of the English Church, with a vindication of the reception by the wicked and of the Adoration of our Lord Jesus Christ truly present. By the Rev. E. B. PUSEY, D.D. 8vo., 6s.

THE LATE REV. J. KEBLE, M.A.

STUDIA SACRA. COMMENTARIES on the Introductory Verses of St. John's Gospel, and on a Portion of St. Paul's Epistle to the Romans; with other Theological Papers by the late Rev. JOHN KEBLE, M.A. 8vo., cl., 10s. 6d.

OCCASIONAL PAPERS AND REVIEWS. By the late Rev. JOHN KEBLE, Author of "The Christian Year." Demy 8vo., cloth extra, 12s.

LETTERS OF SPIRITUAL COUNSEL AND GUIDANCE. By the late Rev. J. KEBLE, M.A., Vicar of Hursley. Edited, with a New Preface, by R. F. WILSON, M.A., Vicar of Rownhams, &c. Third Edition, much enlarged, Post 8vo., cloth, 6s.

OUTLINES OF INSTRUCTIONS OR MEDITATIONS FOR THE CHURCH'S SEASONS. By JOHN KEBLE, M.A. Edited, with a Preface, by R. F. WILSON, M.A. Crown 8vo., toned paper, cloth, 5s.

THE LATE BISHOP OF BRECHIN.

AN EXPLANATION OF THE THIRTY-NINE ARTICLES. With an Epistle Dedicatory to the Rev. E. B. PUSEY, D.D. By A. P. FORBES, D.C.L., Bishop of Brechin. Second Edition, Crown 8vo., cloth, 12s.

A SHORT EXPLANATION OF THE NICENE CREED, for the Use of Persons beginning the Study of Theology. By ALEXANDER PENROSE FORBES, D.C.L., Bishop of Brechin. *Second Edition.* Crown 8vo., cloth, 6s.

THE LORD BISHOP OF SALISBURY.

THE ADMINISTRATION OF THE HOLY SPIRIT IN THE BODY OF CHRIST. The Bampton Lectures for 1868. By GEORGE MOBERLY, D.C.L., Lord Bishop of Salisbury. *2nd Edit.* Crown 8vo., cloth, 7s. 6d.

SERMONS ON THE BEATITUDES, with others mostly preached before the University of Oxford. By the same. *Third Edition.* Crown 8vo., cloth, 7s. 6d.

REV. WILLIAM BRIGHT, D.D.

A HISTORY OF THE CHURCH, from the Edict of Milan, A.D. 313, to the Council of Chalcedon, A.D. 451. *Second Edition.* Post 8vo., 10s. 6d.

THEOLOGICAL WORKS, &c. (continued).

APOLLOS; or, THE WAY OF GOD. A Plea for the Religion of Scripture. By A. CLEVELAND COXE, Bishop of New York. Crown 8vo., cl., 5s.

THE HISTORY OF CONFIRMATION. By WILLIAM JACKSON, M.A., Queen's College, Oxford; Vicar of Heathfield, Sussex. Crown 8vo., cloth, 4s.

A COMMENTARY ON THE EPISTLES AND GOSPELS IN THE BOOK OF COMMON PRAYER. Extracted from Writings of the Fathers of the Holy Catholic Church, anterior to the Division of the East and West. With an Introductory Notice by the DEAN OF ST. PAUL'S. In Two Vols., Crown 8vo., cloth, 15s.

THE EXPLANATION OF THE APOCALYPSE by VENERABLE BEDA, Translated by the Rev. EDW. MARSHALL, M.A., F.S.A., formerly Fellow of Corpus Christi College, Oxford. 180 pp. Fcap. 8vo., cloth, 3s. 6d.

GODET'S BIBLICAL STUDIES ON THE OLD TESTAMENT. Edited by the Hon. and Rev. W. H. LYTTELTON. Fcap. 8vo., cloth, 6s.

THE CATHOLIC DOCTRINE OF THE SACRIFICE AND PARTICIPATION OF THE HOLY EUCHARIST. By GEORGE TREVOR, D.D., M.A., Canon of York; Rector of Beeford. Second Edition, revised and enlarged. Crown 8vo., cloth, 10s. 6d.

THE LAST TWELVE VERSES OF THE GOSPEL ACCORDING TO S. MARK Vindicated against Recent Critical Objectors and Established, by JOHN W. BURGON, B.D., Dean of Chichester. With Facsimiles of Codex א and Codex L. 8vo., cloth, 12s.

DISCOURSES ON PROPHECY. In which are considered its Structure, Use, and Inspiration. By JOHN DAVISON, B.D. A New Edition. 8vo., cloth, 9s.

THE PRINCIPLES OF DIVINE SERVICE; or, An Inquiry concerning the True Manner of Understanding and Using the Order for Morning and Evening Prayer, and for the Administration of the Holy Communion in the English Church. By the late Ven. PHILIP FREEMAN, Archdeacon of Exeter. *A New Edition.* 2 vols., 8vo., cloth, 16s.

CATENA AUREA. A Commentary on the Four Gospels, collected out of the Works of the Fathers by S. THOMAS AQUINAS. Uniform with the Library of the Fathers. Re-issue. Complete in 6 vols. 8vo., cloth, £2 2s.

CHRISTIANITY AS TAUGHT BY S. PAUL. The Bampton Lectures for 1870. To which is added an Appendix of the Continuous Sense of S. Paul's Epistles; with Notes and Metalegomena. By the Rev. W. J. IRONS, D.D., &c. Second Edition, with New Preface, 8vo., with Map, cloth, 9s.

CHARACTERISTICS OF CHRISTIAN MORALITY. The Bampton Lectures for 1873. By the Rev. I. GREGORY SMITH, M.A. Second Edition, Crown 8vo., cloth, 3s. 6d.

BEDE'S ECCLESIASTICAL HISTORY OF THE ENGLISH NATION. A New Translation by the Rev. L. GIDLEY, M.A., Chaplain of St. Nicholas', Salisbury. Crown 8vo., cloth, 6s.

RICHARD BAXTER ON THE SACRAMENTS: Holy Orders, Holy Baptism, Confirmation, Absolution, Holy Communion. 18mo., cloth, 1s.

THE CONSTITUTIONS AND CANONS ECCLESIASTICAL OF THE CHURCH OF ENGLAND, Referred to their Original Sources, and Illustrated with Explanatory Notes. By the late MACKENZIE E. C. WALCOTT, B.D., F.S.A. Fcap. 8vo., cloth, 4s.

A CRITICAL HISTORY OF THE ATHANASIAN CREED, by the Rev. D. WATERLAND, D.D. Edited by the Rev. J. R. KING, M.A. Fcap. 8vo., cloth, 5s.

THE PASTORAL RULE OF ST. GREGORY. Sancti Gregorii Papæ Regulæ Pastoralis Liber, ad Johannem Episcopum Civitatis Ravennæ. With an English Translation. By the Rev. H. R. BRAMLEY, M.A., Fellow of Magdalen College, Oxford. Fcap. 8vo., cloth, 6s.

THE DEFINITIONS OF THE CATHOLIC FAITH and Canons of Discipline of the first four General Councils of the Universal Church. In Greek and English. Fcap. 8vo., cloth, 2s. 6d.

DE FIDE ET SYMBOLO: Documenta quædam nec non Aliquorum SS. Patrum Tractatus. Edidit CAROLUS A. HEURTLEY, S.T.P., Dom. Margaretæ Prælector, et Ædis Christi Canonicus. Fcap. 8vo., cloth, 4s. 6d.

S. AURELIUS AUGUSTINUS, Episcopus Hipponensis, de Catechizandis Rudibus, de Fide Rerum quæ non videntur, de Utilitate Credendi. In Usum Juniorum. Edidit C. MARRIOTT, S.T.B., Olim Coll. Oriel. Socius. *New Edition.* Fcap. 8vo., cloth, 3s. 6d.

ANALECTA CHRISTIANA, In usum Tironum. Excerpta, Epistolæ, &c., ex EUSEBII, &c.; S. IGNATII Epistolæ ad Smyrnæos et ad Polycarpum; E. S. CLEMENTIS ALEXANDRI Pædagogo excerpta; S. ATHANASII Sermo contra Gentes. Edidit et Annotationibus illustravit C. MARRIOTT, S.T.B. 8vo., 10s. 6d.

S. PATRIS NOSTRI S. ATHANASII ARCHIEPISCOPI ALEXANDRIÆ DE INCARNATIONE VERBI, ejusque Corporali ad nos Adventu. With an English Translation by the Rev. J. RIDGWAY, B.D., Hon. Canon of Christ Church. Fcap. 8vo., cloth, 5s.

OXFORD SERIES OF DEVOTIONAL WORKS.

Fcap. 8vo., printed in Red and Black, on toned paper.

The Imitation of Christ.
FOUR BOOKS. By Thomas A KEMPIS. Cloth, 4s.—Pocket Edit., 32mo., cl., 1s.

Andrewes' Devotions.
DEVOTIONS. By the Right Rev. Father in God, LAUNCELOT ANDREWES. Translated from the Greek and Latin, and arranged anew. Antique cloth, 5s.

Taylor's Holy Living.
THE RULE AND EXERCISES OF HOLY LIVING. By BISHOP JEREMY TAYLOR. Antique cloth, 4s.—Pocket Edition, 32mo., cloth, 1s.

Taylor's Holy Dying.
THE RULE AND EXERCISES OF HOLY DYING. By BISHOP JEREMY TAYLOR. Antique cloth, 4s.—Pocket Edition, 32mo., cloth, 1s.

Taylor's Golden Grove.
THE GOLDEN GROVE; a Choice Manual, containing what is to be Believed, Practised, and Desired, or Prayed for. By BISHOP JEREMY TAYLOR. Printed uniform with "Holy Living and Holy Dying." Antique cloth, 3s. 6d.

Sutton's Meditations.
GODLY MEDITATIONS UPON THE MOST HOLY SACRAMENT OF THE LORD'S SUPPER. By CHRISTOPHER SUTTON, D.D., late Prebend of Westminster. A new Edition. Antique cloth, 5s.

Laud's Devotions.
THE PRIVATE DEVOTIONS of DR. WILLIAM LAUD, Archbishop of Canterbury, and Martyr. Antique cloth, 5s.

Spinckes' Devotions.
TRUE CHURCH OF ENGLAND MAN'S COMPANION IN THE CLOSET; or, a complete Manual of Private Devotions, collected from the Writings of eminent Divines of the Church of England. Antique cloth, 4s.

Ancient Collects.
ANCIENT COLLECTS AND OTHER PRAYERS. Selected for Devotional use from various Rituals. By WM. BRIGHT, D.D. Antique cloth, 5s.

Devout Communicant.
THE DEVOUT COMMUNICANT, exemplified in his Behaviour before, at, and after the Sacrament of the Lord's Supper: Practically suited to all the Parts of that Solemn Ordinance. 7th Edition, revised. Fcap. 8vo., cloth, 4s.

ΕΙΚΩΝ ΒΑΣΙΛΙΚΗ.
THE PORTRAITURE OF HIS SACRED MAJESTY KING CHARLES I. in his Solitudes and Sufferings. New Edition, with Preface by C. M. PHILLIMORE [On the evidence that the book was written by Charles I., and not by Gauden]. Antique cloth, 5s.

DEVOTIONAL.

THE LIFE OF JESUS CHRIST IN GLORY: Daily Meditations, from Easter Day to the Wednesday after Trinity Sunday. By NOUET. Translated from the French, and adapted to the Use of the English Church. *Third Thousand.* 12mo., cloth, 5s.

A GUIDE FOR PASSING ADVENT HOLILY. By AVRILLON. Translated from the French, and adapted to the use of the English Church. *New Edition.* Fcap. 8vo., cloth, 5s.

A GUIDE FOR PASSING LENT HOLILY. By AVRILLON. Translated from the French, and adapted to the use of the English Church. Fourth Edition. Fcap. 8vo., cloth, 5s.

THE PASTOR IN HIS CLOSET; or, A Help to the Devotions of the Clergy. By JOHN ARMSTRONG, D.D., late Lord Bishop of Grahamstown. *Third Edition.* Fcap. 8vo., cloth, 2s.

DAILY STEPS TOWARDS HEAVEN; or, Practical Thoughts on the Gospel History, for every day in the year. With Titles and Characters of Christ. 32mo., roan, 2s. 6d. Large type edition, Crown 8vo., cloth, 5s.

Uniform with above.

THE HOURS; being Prayers for the Third, Sixth, and Ninth Hours; with a Preface, and Heads of Devotion for the Day. Seventh Edition. 32mo., in parchment wrapper, 1s.

ANNUS DOMINI. A Prayer for each Day of the Year, founded on a Text of Holy Scripture. By CHRISTINA G. ROSSETTI. 32mo., cloth, 3s. 6d.

DEVOTIONS BEFORE AND AFTER HOLY COMMUNION. With Prefatory Note by KEBLE. Sixth Edition, in red and black, on toned paper, 32mo., cloth, 2s.—With the Service, 32mo., cloth, 2s. 6d.

MEDITATIONS FOR THE FORTY DAYS OF LENT. With a Prefatory Notice by the ARCHBISHOP OF DUBLIN. 18mo., cloth, 2s. 6d.

THE EVERY-DAY COMPANION. By the Rev. W. H. RIDLEY, M.A., Rector of Hambleden, Bucks. Fcap. 8vo., cloth, 3s.

EVENING WORDS. Brief Meditations on the Introductory Portion of our Lord's Last Discourse with His Disciples. 16mo., cloth, 2s.

THOUGHTS DURING SICKNESS. By ROBERT BRETT, Author of "The Doctrine of the Cross," &c. Fcap. 8vo., limp cloth, 1s. 6d.

BREVIATES FROM HOLY SCRIPTURE, arranged for use by the Bed of Sickness. By the Rev. G. ARDEN, M.A., Rector of Winterborne-Came; Domestic Chaplain to the Right Hon. the Earl of Devon. *2nd Ed.* Fcap. 8vo., 2s.

DEVOTIONS FOR A TIME OF RETIREMENT AND PRAYER FOR THE CLERGY. New Edition, revised. Fcap. 8vo., cloth, 1s.

PRAYERS IN USE AT CUDDESDON COLLEGE. Third Edition, revised and enlarged. Fcap. 8vo., 2s.

EARL NELSON'S FAMILY PRAYERS. With Responsions and Variations for the different Seasons, for General Use. New and improved Edition, *large type*, cloth, 2s.

INSTRUCTIONS ON THE HOLY EUCHARIST, AND DEVOTIONS FOR HOLY COMMUNION, being Part V. of the Clewer Manuals, by Rev. T. T. CARTER, M.A., Rector of Clewer. 18mo., cloth, 2s.

THE SERVICE-BOOK OF THE CHURCH OF ENGLAND, arranged according to the New Table of Lessons. Crown 8vo., roan, 12s.; calf antique or calf limp, 16s.; limp morocco or best morocco, 18s.

SERMONS, &c.

PAROCHIAL SERMONS. By E. B. Pusey, D.D. Vol. I. From Advent to Whitsuntide. *Seventh Edition.* 8vo., cloth, 6s. Vol. II., 8vo., cl., 6s.
————— Vol. III. (Reprinted from "Plain Sermons by Contributors to Tracts for the Times.") Revised Edition. 8vo., cloth, 6s.
————— Vol. IV. [*Shortly.*

PAROCHIAL SERMONS preached and printed on Various Occasions, 1832–1850. By E. B. Pusey, D.D. 8vo., cloth, 6s.

SERMONS preached before the UNIVERSITY OF OXFORD between A.D. 1859 and 1872. Vol. II. By E. B. Pusey, D.D. 8vo., cloth, 6s.
————— 1843 to 1855, &c. Vol. I. 8vo., cloth, 6s.
————— 1864 to 1876, &c. Vol. III. 8vo., cloth, 6s.

LENTEN SERMONS preached chiefly to Young Men at the Universities, between A.D. 1868 and 1874. By E. B. Pusey, D.D. 8vo., cloth, 6s.

ILLUSTRATIONS OF FAITH. Eight Plain Sermons, by the late Rev. Edward Monro. Fcap. 8vo., cloth, 2s. 6d.

Uniform, and by the same Author,

PLAIN SERMONS ON THE BOOK OF COMMON PRAYER. Fcap. 8vo., cloth, 5s.
SERMONS ON NEW TESTAMENT CHARACTERS. Fcap. 8vo., 4s.
HISTORICAL AND PRACTICAL SERMONS ON THE SUFFERINGS AND RESURRECTION OF OUR LORD. 2 vols., Fcap. 8vo., cloth, 10s.

CHRISTIAN SEASONS.—Short and Plain Sermons for every Sunday and Holyday throughout the Year. 4 vols., Fcap. 8vo., cloth, 10s. Second Series, 4 vols., Fcap. 8vo., cloth, 10s.

SHORT SERMONS FOR FAMILY READING, following the Order of the Christian Seasons. By the Rev. J. W. Burgon, B.D. 2 vols., Fcap. 8vo., cloth, 8s. Second Series, 2 vols., Fcap. 8vo., cloth, 8s.

PAROCHIAL SERMONS. By the late Bp. Armstrong. Fcap. 8vo., cloth, 5s.

SERMONS FOR FASTS AND FESTIVALS. By the late Bp. Armstrong. A New Edition. Fcap. 8vo., 5s.

SERMONS FOR THE CHRISTIAN YEAR. By J. Keble, M.A.
ADVENT TO CHRISTMAS. 8vo., cl., 6s.
CHRISTMAS AND EPIPHANY. 8vo., cloth, 6s.
SEPTUAGESIMA TO LENT. 8vo., cl., 6s.
ASH-WEDNESDAY TO HOLY WEEK. 8vo., cloth, 6s.
HOLY WEEK. 8vo., cloth, 6s.
EASTER TO ASCENSION DAY. 8vo., cloth, 6s.
ASCENSION DAY TO TRINITY SUNDAY inclusive. 8vo., cloth, 6s.
TRINITY, Part I. 8vo., cloth, 6s.
TRINITY, Part II. 8vo., cloth, 6s.
SAINTS' DAYS. 8vo., cloth, 6s.

VILLAGE SERMONS ON THE BAPTISMAL SERVICE. By the Rev. John Keble, M.A. 8vo., cloth, 5s.

THE AWAKING SOUL, as Sketched in the 130th Psalm. Addresses delivered in Lent, 1877. By E. R. Wilberforce, M.A. Crown 8vo., limp cloth, 2s. 6d.

OXFORD LENT SERMONS, 1857, 8, 9, 60, 3, 5, 6, 7, 8, 9, 70—71. cloth, 5s. each.

XX. SHORT ALLEGORICAL SERMONS. By B. K. W. Pearse, M.A., and W. A. Gray, M.A. *Sixth Edition,* Fcap. 8vo., sewed, 1s.

SERMONS AND ESSAYS ON THE APOSTOLICAL AGE. By the Very Rev. Arthur Penrhyn Stanley, D.D. *Third Edition, revised.* Crown 8vo., cloth, 7s. 6d.

Works of the Standard English Divines,

PUBLISHED IN THE LIBRARY OF ANGLO-CATHOLIC THEOLOGY,

AT THE FOLLOWING PRICES IN CLOTH.

ANDREWES' (BP.) COMPLETE WORKS. 11 vols., 8vo., £3 7s.
    THE SERMONS. (Separate.) 5 vols., £1 15s.

BEVERIDGE'S (BP.) COMPLETE WORKS. 12 vols., 8vo., £4 4s.
    THE ENGLISH THEOLOGICAL WORKS. 10 vols., £3 10s.

BRAMHALL'S (ABP.) WORKS, WITH LIFE AND LETTERS, &c. 5 vols., 8vo., £1 15s. (Vol. 2 cannot be sold separately.)

BULL'S (BP.) HARMONY ON JUSTIFICATION. 2 vols., 8vo., 10s.
———————— DEFENCE OF THE NICENE CREED. 2 vols., 10s.
———————— JUDGMENT OF THE CATHOLIC CHURCH. 5s.

COSIN'S (BP.) WORKS COMPLETE. 5 vols., 8vo., £1 10s.

CRAKANTHORP'S DEFENSIO ECCLESIÆ ANGLICANÆ. 8vo., 7s.

FRANK'S SERMONS. 2 vols., 8vo., 10s.

FORBES' CONSIDERATIONES MODESTÆ. 2 vols., 8vo., 12s.

GUNNING'S PASCHAL, OR LENT FAST. 8vo., 6s.

HAMMOND'S PRACTICAL CATECHISM. 8vo., 5s.
———————— MISCELLANEOUS THEOLOGICAL WORKS. 5s.
———————— THIRTY-ONE SERMONS. 2 Parts. 10s.

HICKES'S TWO TREATISES ON THE CHRISTIAN PRIESTHOOD. 3 vols., 8vo., 15s.

JOHNSON'S (JOHN) THEOLOGICAL WORKS. 2 vols., 8vo., 10s.
———————— ENGLISH CANONS. 2 vols., 12s.

LAUD'S (ABP.) COMPLETE WORKS. 7 vols., (9 Parts,) 8vo., £2 17s.

L'ESTRANGE'S ALLIANCE OF DIVINE OFFICES. 8vo., 6s.

MARSHALL'S PENITENTIAL DISCIPLINE. (This volume cannot be sold separate from the complete set.)

NICHOLSON'S (BP.) EXPOSITION OF THE CATECHISM. (This volume cannot be sold separate from the complete set.)

OVERALL'S (BP.) CONVOCATION-BOOK OF 1606. 8vo., 5s.

PEARSON'S (BP.) VINDICIÆ EPISTOLARUM S. IGNATII. 2 vols. 8vo., 10s.

THORNDIKE'S (HERBERT) THEOLOGICAL WORKS COMPLETE. 6 vols., (10 Parts,) 8vo., £2 10s.

WILSON'S (BP.) WORKS COMPLETE. With LIFE, by Rev. J. KEBLE. 7 vols., (8 Parts,) 8vo., £3 3s.

A complete set, 80 *Vols.* in 88 *Parts*, £21.

THE AUTHORIZED EDITIONS OF
THE CHRISTIAN YEAR,
With the Author's latest Corrections and Additions.

NOTICE.—Messrs. PARKER are the sole Publishers of the Editions of the "Christian Year" issued with the sanction and under the direction of the Author's representatives. All Editions without their imprint are unauthorized.

SMALL 4to. EDITION.		32mo. EDITION.	
Handsomely printed on toned paper, with red border lines and initial letters. Cloth extra 10 6		Cloth boards, gilt edges . 1 6	
		Cloth, limp 1 0	
DEMY 8vo. EDITION.		48mo. EDITION.	
Cloth 6 0		Cloth, limp 0 6	
		Roan 1 6	
FOOLSCAP 8vo. EDITION.		FACSIMILE OF THE 1ST EDITION, with a list of the variations from the Original Text which the Author made in later Editions. 2 vols., 12mo., boards . 7 6	
Cloth 3 6			
24mo. EDITION.			
Cloth, red lines . . . 2 6			

The above Editions (except the Facsimile of the First Edition) are kept in a variety of bindings, which may be ordered through the Trade, or direct from the Publishers. The chief bindings are Morocco plain, Morocco Antique, Calf Antique, and Vellum, the prices varying according to the style.

By the same Author.

LYRA INNOCENTIUM. Thoughts in Verse on Christian Children. *Thirteenth Edition.* Fcap. 8vo., cloth, 5s.
—————————— 24mo., cloth, red lines, 3s. 6d.
—————————— 48mo. edition, limp cloth, 6d.; cloth boards, 1s.

MISCELLANEOUS POEMS BY THE REV. JOHN KEBLE, M.A., Vicar of Hursley. [With Preface by G. M.] *Third Edition.* Fcap., cloth, 6s.

THE PSALTER, OR PSALMS OF DAVID: In English Verse. *Fourth Edition.* Fcap., cloth, 6s.
—————————— 18mo., cloth, 1s.

The above may also be had in various bindings.

———

A CONCORDANCE TO THE "CHRISTIAN YEAR." Fcap. 8vo., toned paper, cloth, 4s.

MUSINGS ON THE "CHRISTIAN YEAR;" WITH GLEANINGS FROM THIRTY YEARS' INTERCOURSE WITH THE LATE Rev. J. KEBLE, by CHARLOTTE M. YONGE; to which are added Recollections of Hursley, by FRANCES M. WILBRAHAM. *Second Edition.* Fcap. 8vo., cloth, 7s. 6d.

MEMOIR OF THE REV. J. KEBLE, M.A. By Sir J. T. COLERIDGE. *Fourth and Cheaper Edition.* Post 8vo., cloth, 6s.

THE CHILD'S CHRISTIAN YEAR. Hymns for every Sunday and Holyday throughout the Year. *Cheap Edition,* 18mo., cloth, 1s.

Church Poetry.

RE-ISSUE OF THE POETICAL WORKS OF THE LATE
REV. ISAAC WILLIAMS.

THE CATHEDRAL; or, The Catholic and Apostolic Church in England. 32mo., cloth, 2s. 6d.

THE BAPTISTERY; or, The Way of Eternal Life. With Plates by BOETIUS A BOLSWERT. Fcap. 8vo., cloth, 7s. 6d.; 32mo., cloth, 2s. 6d.

HYMNS FROM THE PARISIAN BREVIARY. 32mo., cloth, 2s. 6d.

THE CHRISTIAN SCHOLAR. Fcap. 8vo., cl., 5s.; 32mo., cl., 2s. 6d.

THOUGHTS IN PAST YEARS. 32mo., cloth, 2s. 6d.

THE SEVEN DAYS OF THE OLD AND NEW CREATION. Fcap. 8vo., cloth, 3s. 6d.

BISHOP CLEVELAND COXE.

CHRISTIAN BALLADS AND POEMS. By ARTHUR CLEVELAND COXE, D.D., Bishop of Western New York. A New Edition. Fcap. 8vo., cloth, 3s. Also selected Poems in a packet, 32mo., 1s.

THE BELLS OF BOTTEVILLE TOWER; A Christmas Story in Verse: and other Poems. By FREDERICK G. LEE, Author of "The Martyrs of Vienne and Lyons," "Petronilla," &c. Fcap. 8vo., with Illustrations, cloth, 4s. 6d.

HYMNS ON THE LITANY, by ADA CAMBRIDGE. Fcap. 8vo., cl., 3s.

Parochial.

THE CONFIRMATION CLASS-BOOK: Notes for Lessons, with APPENDIX, containing Questions and Summaries for the Use of the Candidates. By E. M. HOLMES, LL.B., Rector of Marsh Gibbon, Bucks. Fcap. 8vo., limp cloth, 2s. 6d.

Also, in wrapper, THE QUESTIONS AND SUMMARIES separate, 4 sets of 128 pp. in packet, 1s. each.

THE CATECHIST'S MANUAL; with an Introduction by the late SAMUEL WILBERFORCE, D.D., Lord Bishop of Winchester. By the same. *Sixth Thousand, revised.* Crown 8vo., limp cloth, 5s.

A MANUAL OF PASTORAL VISITATION, intended for the Use of the Clergy in their Visitation of the Sick and Afflicted. By a PARISH PRIEST. Dedicated, by permission, to His Grace the Archbishop of Dublin. Second Edition, Crown 8vo., limp cloth, 3s. 6d.; roan, 4s.

THE INNER LIFE. Hymns on the "Imitation of Christ," by THOMAS A'KEMPIS; designed especially for Use at Holy Communion. By the Author of "Thoughts from a Girl's Life," &c. Fcap. 8vo., cloth, 3s.

SHORT READINGS FOR SUNDAY. By the Author of "Footprints in the Wilderness." With Twelve Illustrations on Wood. Third Thousand. Square Crown 8vo., cloth, 3s. 6d.

A SERIES OF WALL PICTURES illustrating the New Testament. The Set of 16 Pictures, size 22 inches by 19 inches, 12s.

COTTAGE PICTURES FROM THE OLD TESTAMENT. A Series of Twenty-eight large folio Engravings, brilliantly coloured by hand. The Set, 7s. 6d.

COTTAGE PICTURES FROM THE NEW TESTAMENT. A Series of Twenty-eight large folio Engravings, brilliantly coloured. The Set, 7s. 6d.

Upwards of 8,000 Sets of these Cottage Pictures have been sold.

TWELVE SACRED PRINTS FOR PAROCHIAL USE. Printed in Sepia, with Ornamental Borders. *The Set*, One Shilling; or *each*, One Penny.

Upwards of 100,000 of these Prints have already been sold.

FABER'S STORIES FROM THE OLD TESTAMENT. With Four Illustrations. Square Crown 8vo., cloth, 4s.

MUSINGS ON THE PSALM (CXIX.) OF DIVINE ASPIRATIONS. 32mo., cloth, 2s.

R. GODFREY FAUSSETT, M.A.

THE SYMMETRY OF TIME: being an Outline of Biblical Chronology adapted to a Continuous Succession of Weeks of Years. From the Creation of Adam to the Exodus. By R. GODFREY FAUSSETT, M.A., Student of Christ Church. Demy 4to., in wrapper, 10s. 6d.

THE ELEMENTS OF PSYCHOLOGY.

THE ELEMENTS OF PSYCHOLOGY, ON THE PRINCIPLES OF BENEKE, Stated and Illustrated in a Simple and Popular Manner by DR. G. RAUE, Professor in the Medical College, Philadelphia; Fourth Edition, considerably Altered, Improved, and Enlarged, by JOHANN GOTTLIEB DRESSLER, late Director of the Normal School at Bautzen. Translated from the German. Post 8vo., cloth, 6s.

REV. CANON GREGORY.

ARE WE BETTER THAN OUR FATHERS? or, A Comparative View of the Social Position of England at the Revolution of 1688, and at the Present Time. FOUR LECTURES delivered in St. Paul's Cathedral. By ROBERT GREGORY, M.A., Canon of St. Paul's. Crown 8vo., 2s. 6d.

PROFESSOR GOLDWIN SMITH.

THE REORGANIZATION OF THE UNIVERSITY OF OXFORD. By GOLDWIN SMITH. Post 8vo., limp cloth, 2s.

LECTURES ON THE STUDY OF HISTORY. Delivered in Oxford, 1859—61. *Second Edition.* Crown 8vo., limp cloth, 3s. 6d.

IRISH HISTORY AND IRISH CHARACTER. Cheap Edition, Fcap. 8vo., sewed, 1s. 6d.

THE EMPIRE. A Series of Letters published in "The Daily News," 1862, 1863. Post 8vo., cloth, price 6s.

MRS. ALGERNON KINGSFORD.

ROSAMUNDA THE PRINCESS: An Historical Romance of the Sixth Century; the CROCUS, WATER-REED, ROSE and MARIGOLD, PAINTER OF VENICE, NOBLE LOVE, ROMANCE of a RING, and other Tales. By Mrs. ALGERNON KINGSFORD. 8vo., cloth, with Twenty-four Illustrations, 6s.

THE EXILE FROM PARADISE.

THE EXILE FROM PARADISE, translated by the Author of the "Life of S. Teresa." Fcap., cloth, 1s. 6d.

VILHELM THOMSEN.

THE RELATIONS BETWEEN ANCIENT RUSSIA AND SCANDINAVIA, and the Origin of the Russian State. THREE LECTURES delivered at the Taylor Institution, Oxford, in May, 1876, by Dr. VILHELM THOMSEN, Professor at the University of Copenhagen. Small 8vo., cloth, 3s. 6d.

BERNARD BOSANQUET, M.A.

ATHENIAN CONSTITUTIONAL HISTORY, as Represented in Grote's "History of Greece," critically examined by G. F. SCHÖMANN: Translated, with the Author's permission, by BERNARD BOSANQUET, M.A., Fellow and Tutor of University College, Oxford. 8vo., cloth, 3s. 6d.

THE PRAYER-BOOK CALENDAR.

THE CALENDAR OF THE PRAYER-BOOK ILLUSTRATED. (Comprising the first portion of the "Calendar of the Anglican Church," with additional Illustrations, an Appendix on Emblems, &c.) With Two Hundred Engravings from Medieval Works of Art. *Sixth Thousand.* Fcap. 8vo., cl., 6s.

SIR G. G. SCOTT, F.S.A.

GLEANINGS FROM WESTMINSTER ABBEY. By SIR GEORGE GILBERT SCOTT, R.A., F.S.A. With Appendices supplying Further Particulars, and completing the History of the Abbey Buildings, by Several Writers. *Second Edition*, enlarged, containing many new Illustrations by O. Jewitt and others. Medium 8vo., 10s. 6d.

THE LATE CHARLES WINSTON.

AN INQUIRY INTO THE DIFFERENCE OF STYLE OBSERVABLE IN ANCIENT GLASS PAINTINGS, especially in England, with Hints on Glass Painting, by the late CHARLES WINSTON. With Corrections and Additions by the Author. 2 vols., Medium 8vo., cloth, £1 11s. 6d.

REV. SAMUEL LYSONS, F.S.A.

OUR BRITISH ANCESTORS: WHO AND WHAT WERE THEY? An Inquiry serving to elucidate the Traditional History of the Early Britons by means of recent Excavations, Etymology, Remnants of Religious Worship, Inscriptions, Craniology, and Fragmentary Collateral History. By the Rev. SAMUEL LYSONS, M.A., F.S.A., Rector of Rodmarton, and Perpetual Curate of St. Luke's, Gloucester. Post 8vo., cloth, 5s.

M. VIOLLET-LE-DUC.

THE MILITARY ARCHITECTURE OF THE MIDDLE AGES. Translated from the French of M. VIOLLET-LE-DUC, by M. MACDERMOTT, Esq., Architect. With 151 original French Engravings. Second Edition, with a Preface by J. H. PARKER, C.B. Medium 8vo., cloth, 10s. 6d.

JOHN HEWITT.

ANCIENT ARMOUR AND WEAPONS IN EUROPE. By JOHN HEWITT, Member of the Archæological Institute of Great Britain. Vols. II. and III., comprising the Period from the Fourteenth to the Seventeenth Century, completing the work, £1 12s. Also Vol. I., from the Iron Period of the Northern Nations to the end of the Thirteenth Century, 18s. The work complete, 3 vols., 8vo., £1 11s. 6d.

REV. PROFESSOR STUBBS.

THE TRACT "DE INVENTIONE SANCTÆ CRUCIS NOSTRÆ IN MONTE ACUTO ET DE DUCTIONE EJUSDEM APUD WALTHAM," now first printed from the Manuscript in the British Museum, with Introduction and Notes by WILLIAM STUBBS, M.A. Royal 8vo., 5s.; Demy 8vo., 3s. 6d.

NORTHERN ANTIQUITIES.

THE PRIMEVAL ANTIQUITIES of ENGLAND and DENMARK COMPARED. By J. J. A. WORSAAE. Translated and applied to the illustration of similar remains in England, by W. J. THOMS, F.S.A., &c. With numerous Illustrations. 8vo., cloth, 5s.

OUR ENGLISH HOME:

Its Early History and Progress. With Notes on the Introduction of Domestic Inventions. New Edition, Crown 8vo., cloth, 3s. 6d.

PARISH CHURCH GOODS IN BERKSHIRE, A.D. 1552.

Inventories of Furniture and Ornaments remaining in certain of the Parish Churches of Berks in the last year of the reign of King Edward the Sixth: Transcribed from the Original Records, with Introduction and Explanatory Notes by WALTER MONEY, F.S.A., Member of Council for Berks, Brit. Arch. Assoc., and Hon. Sec. of the Newbury District Field Club. Crown 8vo., limp cloth, 3s. 6d.

ARCHITECTURE AND ARCHÆOLOGY.

JOHN HENRY PARKER, C.B., F.S.A., HON. M.A. OXON.

AN INTRODUCTION TO THE STUDY OF GOTHIC ARCHITECTURE. *Fifth Edition*, Revised and Enlarged, with 189 Illustrations, with a Topographical and Glossarial Index. Fcap. 8vo., cloth, 5s.

A CONCISE GLOSSARY OF TERMS USED IN GRECIAN, ROMAN, ITALIAN, AND GOTHIC ARCHITECTURE. A New Edition, revised. Fcap. 8vo., with 470 Illustrations, in ornamental cloth, 7s. 6d.

AN ATTEMPT TO DISCRIMINATE THE STYLES OF ARCHITECTURE IN ENGLAND, from the Conquest to the Reformation; with a Sketch of the Grecian and Roman Orders. By the late THOMAS RICKMAN, F.S.A. *Seventh Edition*, with considerable Additions, chiefly Historical, by JOHN HENRY PARKER, C.B., F.S.A., &c. 8vo. [*Just ready.*

DOMESTIC ARCHITECTURE OF THE MIDDLE AGES, with numerous Engravings from Existing Remains, and Historical Illustrations from Contemporary Manuscripts. By the late T. HUDSON TURNER, Esq. From the Norman Conquest to the Thirteenth Century; interspersed with Remarks on Domestic Manners during the same Period. 8vo., cloth, £1 1s. *A Reprint.*

———— FROM EDWARD I. TO RICHARD II. (the Edwardian Period, or the Decorated Style). By the Editor of "The Glossary of Architecture." 8vo., cloth, £1 1s.

Also,

———— FROM RICHARD II. TO HENRY VIII. (or the Perpendicular Style). With numerous Illustrations of Existing Remains, from Original Drawings. In Two Vols., 8vo., £1 10s.

THE ARCHÆOLOGY OF ROME. With Plates, Plans, and Diagrams. By JOHN HENRY PARKER, C.B.

Part 1. PRIMITIVE FORTIFICATIONS. *Second Edition*, 8vo., with 59 Plates, cloth, 21s.

Part 2. WALLS AND GATES. *Second Edition, nearly ready.*

Part 3. CONSTRUCTION OF WALLS. *Second Edit., in the Press.*

Part 4. THE EGYPTIAN OBELISKS. *Second Edition*, 8vo., cl., 5s.

Part 5. THE FORUM ROMANUM ET MAGNUM. *Second Edit.,* Revised and Enlarged, 41 Plates. 8vo., cloth, 10s. 6d.

Part 6. THE VIA SACRA, and THE TEMPLE OF ROMA, and THE MARBLE PLAN OF ROME originally under the Porticus of that Temple, with 23 Plates.—Also an account of the Excavations in Rome from 1860 to the Present Time.

Part 7. THE COLOSSEUM. 8vo., cloth, 10s. 6d.

Part 8. THE AQUEDUCTS OF ANCIENT ROME. 8vo., cl., 15s.

Part 9. TOMBS IN AND NEAR ROME, and 10. FUNEREAL AND EARLY CHRISTIAN SCULPTURE. 8vo., cloth, 15s.

Part 11. CHURCH AND ALTAR DECORATIONS IN ROME, 8vo., cloth, 10s. 6d.

Part 12. THE CATACOMBS OF ROME. 8vo., cloth, 15s.

Part 13. EARLY AND MEDIÆVAL CASTLES, with an Account of the Excavations in Rome, &c. *Nearly ready.*

Part 14. THE MEDIÆVAL CHURCHES. *Nearly ready.*

SEPULCHRAL CROSSES.

A MANUAL for the STUDY of SEPULCHRAL SLABS and CROSSES of the MIDDLE AGES. By the Rev. EDWARD L. CUTTS, B.A. Illustrated by upwards of 300 Engravings. 8vo., cloth, 6s.

THE ANNALS OF ENGLAND. An Epitome of English History. From Cotemporary Writers, the Rolls of Parliament, and other Public Records. A LIBRARY EDITION, revised and enlarged, with additional Woodcuts. 8vo., half-bound, 12s.

THE SCHOOL EDITION OF THE ANNALS OF ENGLAND. In Five Half-crown Parts. 1. Britons, Romans, Saxons, Normans. 2. The Plantagenets. 3. The Tudors. 4. The Stuarts. 5. The Restoration, to the Death of Queen Anne. Fcap. 8vo., cloth.

THE NEW SCHOOL-HISTORY OF ENGLAND, from Early Writers and the National Records. By the Author of "The Annals of England." Crown 8vo., with Four Maps, limp cloth, 5s.; Coloured Maps, half roan, 6s.

A HISTORY OF THE ENGLISH CHURCH from its Foundation to the Reign of Queen Mary. By M. C. STAPLEY. *Fourth Edition, revised.* Crown 8vo., cloth boards, 5s.

POETARUM SCENICORUM GRÆCORUM, Æschyli, Sophoclis, Euripidis, et Aristophanis, Fabulæ, Superstites, et Perditarum Fragmenta. Ex recognitione GUIL. DINDORFII. Editio Quinta. Royal 8vo., cloth, £1 1s.

THUCYDIDES, with Notes, chiefly Historical and Geographical. By the late T. ARNOLD, D.D. With Indices by the Rev. R. P. G. TIDDEMAN. *Eighth Edition.* 3 vols., 8vo., cloth lettered, £1 16s.

JELF'S GREEK GRAMMAR.—A Grammar of the Greek Language, chiefly from the text of Raphael Kühner. By WM. EDW. JELF, B.D. *Fourth Edition, with Additions and Corrections.* 2 vols. 8vo., £1 10s.

LAWS OF THE GREEK ACCENTS. By JOHN GRIFFITHS, D.D., Warden of Wadham College, Oxford. *Sixteenth Edition.* 16mo., price 6d.

RUDIMENTARY RULES, with Examples, for the Use of Beginners in Greek Prose Composition. By JOHN MITCHINSON, D.C.L., late Head Master of the King's School, Canterbury, (now Bishop of Barbados). 16mo., sewed, 1s.

TWELVE RUDIMENTARY RULES FOR LATIN PROSE COMPOSITION: with Examples and Exercises, for the use of Beginners. By the Rev. E. MOORE, D.D., Principal of St. Edmund Hall, Oxford. *Second Edit.* 16mo., 6d.

MADVIG'S LATIN GRAMMAR. A Latin Grammar for the Use of Schools. By Professor MADVIG, with additions by the Author. Translated by the Rev. G. WOODS, M.A. *New Edition, with an Index of Authors.* 8vo., cloth, 12s.

ERASMI COLLOQUIA SELECTA: Arranged for Translation and Re-translation; adapted for the Use of Boys who have begun the Latin Syntax. By EDWARD C. LOWE, D.D., Canon of Ely, and Provost of the College of SS. Mary and John, Lichfield. Fcap. 8vo., strong binding, 3s.

PORTA LATINA: A Selection from Latin Authors, for Translation and Re-Translation; arranged in a Progressive Course, as an Introduction to the Latin Tongue. By EDWARD C. LOWE, D.D., Editor of Erasmus' "Colloquies," &c. Fcap. 8vo., strongly bound, 3s.

A GRAMMATICAL ANALYSIS OF THE HEBREW PSALTER; being an Explanatory Interpretation of Every Word contained in the Book of Psalms, intended chiefly for the Use of Beginners in the Study of Hebrew. By JOANA JULIA GRESWELL. Post 8vo., cloth, 6s.

SUNDAY-SCHOOL EXERCISES, Collected and Revised from Manuscripts of Burghclere School-children, under the teaching of the Rev. W. B. BARTER, late Rector of Highclere and Burghclere; Edited by his Son-in-law, the BISHOP OF ST. ANDREW'S. *Second Edition.* Crown 8vo., cloth, 5s.

A FIRST LOGIC BOOK, by D. P. CHASE, M.A., Principal of St. Mary Hall, Oxford. Small 4to., sewed, 3s.

NEW AND OLD METHODS OF ETHICS, by F. Y. EDGEWORTH. 8vo., sewed, 3s.

A SERIES OF GREEK AND LATIN CLASSICS
FOR THE USE OF SCHOOLS.

GREEK POETS.

	Cloth. s. d.		Cloth. s. d.
Æschylus	3 0	Sophocles	3 0
Aristophanes. 2 vols.	6 0	Homeri Ilias	3 6
Euripides. 3 vols.	6 6	—— Odyssea	3 0
—— Tragœdiæ Sex	3 6		

GREEK PROSE WRITERS.

Aristotelis Ethica	2 0	Thucydides. 2 vols.	5 0
Demosthenes de Corona, et } Æschines in Ctesiphontem }	2 0	Xenophontis Memorabilia	1 4
		—— Anabasis	2 0
Herodotus. 2 vols.	6 0		

LATIN POETS.

Horatius	2 0	Lucretius	2 0
Juvenalis et Persius	1 6	Phædrus	1 4
Lucanus	2 6	Virgilius	2 6

LATIN PROSE WRITERS.

Cæsaris Commentarii, cum Supplementis Auli Hirtii et aliorum	2 6	Ciceronis Tusc. Disp. Lib. V.	2 0
		Ciceronis Orationes Selectæ	3 6
—— Commentarii de Bello Gallico	1 6	Cornelius Nepos	1 4
		Livius. 4 vols.	6 0
Cicero De Officiis, de Senectute, et de Amicitia	2 0	Sallustius	2 0
		Tacitus. 2 vols.	5 0

TEXTS WITH SHORT NOTES.
UNIFORM WITH THE SERIES OF "OXFORD POCKET CLASSICS."

GREEK WRITERS. TEXTS AND NOTES.
SOPHOCLES.

	s. d.		s. d.
AJAX (*Text and Notes*)	1 0	ANTIGONE (*Text and Notes*)	1 0
ELECTRA ,,	1 0	PHILOCTETES ,,	1 0
ŒDIPUS REX ,,	1 0	TRACHINIÆ ,,	1 0
ŒDIPUS COLONEUS ,,	1 0		

The Notes only, in one vol., cloth, 3s.

ÆSCHYLUS.

PERSÆ (*Text and Notes*)	1 0	CHOEPHORÆ (*Text and Notes*)	1 0
PROMETHEUS VINCTUS ,,	1 0	EUMENIDES ,,	1 0
SEPTEM CONTRA THEBAS ,,	1 0	SUPPLICES ,,	1 0
AGAMEMNON ,,	1 0		

The Notes only, in one vol., cloth, 3s. 6d.

ARISTOPHANES.

THE KNIGHTS (*Text and Notes*) 1 0 | ACHARNIANS (*Text and Notes*) 1 0
THE BIRDS (*Text and Notes*) 1 0

EURIPIDES.

		s.	d.			s.	d.
Hecuba (*Text and Notes*)		1	0	Phœnissæ (*Text and Notes*)		1	0
Medea	,,	1	0	Alcestis ,,		1	0
Orestes	,,	1	0	The above, Notes only, in one vol., cloth, 3s.			
Hippolytus	,,	1	0	Bacchæ ,,		1	0

DEMOSTHENES.

De Corona (*Text and Notes*) . 2 0 | Olynthiac Orations . . 1 0

HOMERUS.

Ilias, Lib. i.—vi. (*Text and Notes*) 2 0

XENOPHON.

Memorabilia (*Text and Notes*) 2 6

ÆSCHINES.

In Ctesiphontem (*Text and Notes*) 2 0

ARISTOTLE.

De Arte Poetica (*Text and Notes*) . cloth, 2s.; sewed 1 6
De Re Publica ,, 3s. ,, 2 6

LATIN WRITERS. TEXTS AND NOTES.

VIRGILIUS.

Bucolica (*Text and Notes*) . 1 0 | Æneidos, Lib. i.—iii. (*Text and Notes*) . . . 1 0
Georgica ,, . 2 0 |

HORATIUS.

Carmina, &c. (*Text and Notes*) 2 0 | Epistolæ et Ars Poetica (*Text and Notes*) . . . 1 0
Satiræ ,, 1 0 |

The Notes only, in one vol., cloth, 2s.

SALLUSTIUS.

Jugurtha (*Text and Notes*) . 1 6 | Catilina (*Text and Notes*) . 1 0

M. T. CICERO.

In Q. Cæcilium — Divinatio (*Text and Notes*) . . 1 0	In Catilinam . . . 1 0
In Verrem Actio Prima . 1 0	Pro Plancio (*Text and Notes*) . 1 6
Pro Lege Manilia, et Pro Archia 1 0	Pro Milone . . . 1 0
	Pro Roscio . . . 1 0
	Orationes Philippicæ, I., II. 1 6

The above, Notes only, in one vol., cloth, 3s. 6d.

De Senectute et De Amicitia 1 0 | Epistolæ Selectæ. Pars I. 1 6

CÆSAR.

De Bello Gallico, Lib. i.—iii. (*Text and Notes*) . . 1 0

CORNELIUS NEPOS.

Lives (*Text and Notes*) . . 1 6

PHÆDRUS.

Fabulæ (*Text and Notes*) . 1 0

LIVIUS.

Lib. xxi.—xxiv. (*Text and Notes*) sewed 4 0
Ditto in cloth . . . 4 6

TACITUS.

The Annals. Notes only, 2 vols., 16mo., cloth 7 0

Portions of several other Authors are in preparation.

Uniform with the Oxford Pocket Classics.

THE LIVES OF THE MOST EMINENT ENGLISH POETS; WITH CRITICAL OBSERVATIONS ON THEIR WORKS. By Samuel Johnson. 3 vols., 24mo., cloth, 2s. 6d. each.

THE LIVES OF MILTON AND POPE, with Critical Observations on their Works. By Samuel Johnson. 24mo., cloth, 1s. 6d.

CHOICE EXTRACTS FROM MODERN FRENCH AUTHORS, for the use of Schools. 18mo., cloth, 3s.

SELECTIONS from the RECORDS of the CITY OF OXFORD, with Extracts from other Documents illustrating the Municipal History: Henry VIII. to Elizabeth [1509—1583]. Edited, by authority of the Corporation of the City of Oxford, by WILLIAM H. TURNER, of the Bodleian Library; under the direction of ROBERT S. HAWKINS, Town Clerk. Royal 8vo., cloth, £1 1s.

A HANDBOOK FOR VISITORS TO OXFORD. Illustrated with numerous Woodcuts by Jewitt, and Steel Plates by Le Keux. *A New Edition.* 8vo., cloth, 12s.

THE OXFORD UNIVERSITY CALENDAR for 1881. Corrected to the end of December, 1880. 12mo., cloth, 4s. 6d.

THE OXFORD TEN-YEAR BOOK: A Complete Register of University Honours and Distinctions, made up to the end of the Year 1870. Crown 8vo., roan, 7s. 6d.

WYKEHAMICA: a History of Winchester College and Commoners, from the Foundation to the Present Day. By the Rev. H. C. ADAMS, M.A., late Fellow of Magdalen College, Oxford. Post 8vo., cloth, 508 pp., with Nineteen Illustrations, 10s. 6d.

HISTORICAL TALES

Illustrating the Chief Events in Ecclesiastical History, British and Foreign.

29 Numbers at One Shilling each, limp cloth; or arranged in Six Volumes, cloth lettered, 3s. 6d. each.

ADDITIONAL VOLUMES TO THE SERIES.

ENGLAND: Mediæval Period. Containing The Orphan of Evesham, or The Jews and the Mendicant Orders.—Mark's Wedding, or Lollardy.—The White Rose of Lynden, or The Monks and the Bible.—The Prior's Ward, or The Broken Unity of the Church. By the Rev. H. C. ADAMS, Vicar of Dry Sandford; Author of "Wilton of Cuthbert's," "Schoolboy Honour," &c. With Four Illustrations on Wood. Fcap. 8vo., cloth, 3s. 6d.

THE ANDREDS-WEALD, OR THE HOUSE OF MICHELHAM: A Tale of the Norman Conquest. By the Rev. A. D. CRAKE, B.A., Fellow of the Royal Historical Society; Author of "Æmilius," "Alfgar the Dane," &c. With Four Illustrations by LOUISA TAYLOR. Fcap. 8vo., cloth, 3s. 6d.

CHEAPER ISSUE OF TALES FOR YOUNG MEN AND WOMEN.

In Six Half-crown Vols., cloth.

Vol. I. contains F. E. PAGET's Mother and Son, Wanted a Wife, and Hobson's Choice.

Vol. II. F. E. PAGET's Windycote Hall, Squitch, Tenants at Tinkers' End.

Vol. III. W. E. HEYGATE's Two Cottages, The Sisters, and Old Jarvis's Will.

Vol. IV. W. E. HEYGATE's James Bright the Shopman, The Politician, Irrevocable.

Vol. V. R. KING's The Strike, and Jonas Clint; N. BROWN's Two to One, and False Honour.

Vol. VI. J. M. NEALE's Railway Accident; E. MONRO's The Recruit, Susan, Servants' Influence, Mary Thomas, or Dissent at Evenly; H. HAYMAN's Caroline Elton, or Vanity and Jealousy.

Each Volume is bound as a distinct and complete work, and sold separately for PRESENTS.

Advertisements.

BEMROSE AND SONS' LIST.

THE CLERGYMAN'S READY REFERENCE REGISTER
Of Services, Occasional Offices, Confirmation, Churchwardens' Accounts, Summaries, Parish Meetings, and Clubs; with Private and Miscellaneous Matter. Arranged on an Original Plan. By the Rev. THEODORE JOHNSON.
A Edition, 230 pp., in boards, 10s. 6d.; or in extra strong binding, with Lock and Key, fitted 14s. 6d.
B Edition, 121 pp., containing the Fourteen Registers, in boards, price 6s. 6d.; or in extra strong binding, with Lock and Key fitted, price 10s. 6d.
Specimen Openings post free on application.

"An extremely convenient book for any one who cherishes the valuable habit of order in his ministrations. It is a very complete register, and any Clergyman who will keep it lying on his desk, and fill in the items as occasion occurs, will find at the end of ten years that he has a valuable statistical epitome of all that has been going on in the decade. And while it is going on it will be found, as Mr. Johnson suggests, a most useful auxiliary in collecting and retaining materials for the monthly parish magazine or the annual almanack."
—*Guardian.*

"Mr. Johnson has performed his task extremely well, and has thus provided a register which is admirably suited to the purpose in view."—*Rock.*

"We can recommend the volume for clerical use."—*Record.*

CHURCH DOOR CALENDARS.
Space for each day in the week for Notices of Services, Lectures, &c. Specimens on application. Size, 16 by 10 inches. The set of 52, 2s. 6d.; headed with name of Parish, 3s. 6d.
⁎ *Oxford Frame for ditto (without Glass) price 2s. 6d.*

SERVICE TABLES (Choir Notices). Printed in red and black, to be filled in with the Chants, Hymn Tunes, &c., for Reading Desk, Choir, Organ, &c. Specimens on application. Eight varieties, 1s. 6d. per 100.
⁎ *Oxford Frame for ditto (without Glass) price 1s. 6d.*

ILLUMINATED LETTERS.
For Church and School Decoration, printed in Two Colours on paper, assorted to order. Size A, 5 inches in height, 1s. per dozen. Size B, 3½ in. in height, 7d. per dozen.

CONFIRMATION CARDS.
Illuminated for Framing. First Quality, 32s. per 100, 5s. per dozen, 6d. each; Second Quality, 14s. per 100, 1s. 8d. per doz., 2d. each. Size, 9 in. by 6 in. Plain ones printed at cheap rates.

BAPTISMAL AND SPONSORS' CARDS.
Printed in two colours, 8s. per 100, 1s. 3d. per doz., 1½d. each.

BANNS BOOK.
200 leaves, bound in Calf, 12s. 6d.; 100 leaves, Calf, 8s. 6d.; 100 leaves, Forrel, 7s. 6d.; 50 leaves, Calf, 6s.; 50 leaves, Forrel, 5s.

BEMROSE'S PREACHER'S BOOK.
Printed and Ruled, for Register of Services, Sermons, Collections, &c. Fcap. Folio. With 60 leaves, price 5s. 6d.; with 100 leaves, 8s.; and with 150 leaves, 10s.

CHURCH DECORATION IN PAPER ROSETTE WORK.
Illustrated by Thirty-five Lithographed Designs, &c. By W. BEMROSE. Price 6d.; post free, 7d.

ECCLESIASTICAL & ACADEMICAL COLOURS.
Part I.—An explanation of the various Colours used in the Services of the Church, with Tables of Colours. Part II.—A List of Hoods worn by Graduates of British and Colonial Universities, and Members of Theological Colleges; with a short Account of those Universities and Colleges which confer Degrees and Grant Hoods. Compiled by the Rev. T. W. WOOD. Fcap. 8vo., cloth, 1s. 6d.

BIBLE CLASS CARDS.
Illuminated for framing. 15s. per 100, 2s. per doz. Name of Class inserted without extra charge if 200 be ordered.

TEMPERANCE PLEDGE CARDS.
Illuminated for framing. 15s. per 100, 2s. per doz. Name of Society inserted without extra charge if 200 be ordered.

MESSRS. BEMROSE & SONS *give special attention to Clerical and Parochial Printing and Publishing. Estimates furnished.*

LONDON: 23, OLD BAILEY, E.C.; DERBY: IRONGATE.

Church Congress Report.

SEVENTH YEAR.   Siempre Adelante.
L'ART:
A WEEKLY ARTISTIC REVIEW
OF
Painting, Sculpture, Architecture, Archæology, The Drama, & Music.

TERMS OF SUBSCRIPTION, &c.

For the United Kingdom: One Year, £5 5s.; Six Months, £2 12s. 6d.; Three Months, £1 7s.
For the United States & Canada: One Year, £6 3s.; Six Months, £3 1s.; Three Months, £1 11s.
For India: One Year, £7 5s.; Six Months, £3 12s. 6d.; Three Months, £1 17s.

All Subscriptions are payable in advance. Subscriptions may commence from 1st January, 1st April, 1st July, and 1st October. No Subscriptions can be received for a less period than three months. Subscribers can, at their option, receive the Journal in quarterly volumes.

Director for England—J. COMYNS CARR, Esq.
Manager—Mr. M. A. JOHNSTON.

The following may be mentioned among the Etchings recently issued in L'ART:—

Venetian Pearl Stringers. After C. C. Van HANNEN; by RAMUS.
Isle-Les-Villenoy: Bards de la Marne. Etched and painted by E. C. YON.
La Fontaine. By J. J. HENNER; etched by Ch. COURTRY.
Portrait de M. Grevy. By L. BONNAT; etched by AD. LALAUZE.
Les Deux Amies. By J. CARAUD; etched by Mme. C. CHOLE-MOUTET.
Les Enerves de Jumieges. E. V. LUMINAIS; etched by GAUJEAN.
Le Martin. BERNIER. Etched by YON.

The Three Graces. After G. F. WATTS, R.A.
Venice. An original etching by J. P. HESELTINE.
Sons of the Brave. After P. R. MORRIS.
Un Coin de Jardin. After CASANOVA; by CHAMPOLLION.
Coast Pastures. After MARK FISHER; by HISON.
Le Halte. After MEISSONNIER; by GALAUZE.
Near Naples. After GALOFRE; by GREVE
Child, Cat, and Dog. After DESBOUTINS.

Etc. Etc. Etc.

There will also be published numerous Engravings, on Wood and in Fac-simile, from works by Sir F. LEIGHTON, L. ALMA TADEMA, J. D. LINTON, A. STEVENS, VAN MARCKE, E. BURNE JONES, W. R. MACBETH, &c., &c., &c.

PRICE OF SINGLE NUMBERS, 2s. each.

Single volumes purchased after publication, £1 15s.; in boards, with gilt tops £2; with gilt edges, £2 15s.

EDITIONS OF SUPERIOR QUALITY.

L'ART publishes two editions of superior quality: the first, limited to 100 copies with the text upon Dutch paper, is accompanied by two series of plates, the one with letters, and the other a proof before letters on Japanese paper; the second, limited to five copies, has four series of plates, viz., upon Dutch paper with letters, on Japanese paper before letters, upon Vellum before letters, upon Whatman before letters. These editions are numbered and the proofs before letters bear the artist's signature.

TERMS OF SUBSCRIPTION.

To the Edition of 100 copies, £16 a year; for the Edition of five copies, £48 a year. Subscriptions for these editions are not received for less than a year.

** The Annual Subscribers to L'ART have the exclusive right to the large presentation plates published every year.

ALBUMS OF PROOF ETCHINGS.

THE CHAUVEL ALBUM: containing Twelve Proof Etchings by Theophili Chauvel.
THE JACQUEMART ALBUM: containing Eight Proof Etchings by the late Jules Jacquemart.
THE WALKER ALBUM: containing Twelve Proof Etchings by Charles Walker.
THE GREUX ALBUM: containing Twenty Proof Etchings by Gustave Greux.

Also a variety of Etchings, elegantly framed, by Rason, Lalauze, Champollion, &c., &c., suitable for presents. Collectors will also find a most varied selection in our Stock of Etchings, by Modern and Ancient Masters, both English and Foreign, of which an inspection is invited.

LIBRAIRIE DE L'ART, 134, NEW BOND STREET.

Advertisements.

A SPLENDID GIFT BOOK.
Just Published, in Two Vols., Imperial 8vo.,
ROMOLA. By GEORGE ELIOT.
WITH ILLUSTRATIONS BY SIR FREDERICK LEIGHTON, PRESIDENT OF THE ROYAL ACADEMY.

The number of Copies Printed is limited to *One Thousand*, each copy being numbered. The mode of publication adopted is that of Subscription *through Booksellers*. Information regarding the Terms of Subscription may be obtained from any Bookseller.

The Thackeray Gift Book.   Illustrated Edition of Thackeray's Ballads.
Small 4to., 16s.
BALLADS. By WILLIAM MAKEPEACE THACKERAY.

With a Portrait of the Author, and 56 Illustrations by the Author, Miss BUTLER (Miss Elizabeth Thompson), GEORGE DU MAURIER, JOHN COLLIER, H. FURNISS, G. G. KILBURNE, M. FITZGERALD, and J. P. ATKINSON. Printed on Toned Paper by CLAY, SONS, & TAYLOR, and elegantly bound in cloth, gilt edges, by BURN.

THE WORKS OF W. M. THACKERAY.
A MAGNIFICENT PRESENT FOR CHRISTMAS OR THE NEW YEAR.
Twenty-four Vols., Imperial 8vo.,

Containing 248 Steel Engravings, 1473 Wood Engravings, and 88 Coloured Illustrations. The Steel and Wood Engravings are all printed on real China Paper, and mounted. Only *One Thousand Copies* printed, each set numbered. The mode of Publication adopted is that of Subscription *through Booksellers*. Terms of Subscription, &c., from any Bookseller.

LONDON: SMITH, ELDER & Co., 15, Waterloo Place.

HERDER'S SCRIPTURE PRINTS.

40 GERMAN PICTURES, ILLUSTRATING THE OLD AND NEW TESTAMENT,

Beautifully Coloured by hand (Published by HERDER, of Friburgh),
ARE VERY SUITABLE FOR SCHOOLS, COTTAGES, &c.

In folio, 12s. the set; or Mounted and Varnished, price £2 2s.

A fresh supply just received.

JOHN HODGES, 24, KING WILLIAM STREET, CHARING CROSS.

EAST GRINSTEAD SCHOOL OF EMBROIDERY,
ST. KATHARINE'S, 32, QUEEN SQUARE, W.C.

Orders are taken for Eucharistic Vestments, Altar Cloths, Alms Bags, Surplices, Cassocks, Banners, Hangings, &c.

Crewel Work designed, traced, and begun for ladies' own working. Materials supplied.

Orders for Plain Needlework are also gratefully received. The Altar Breads supplied, of various sizes and patterns, at the usual prices.

PRICE LISTS ON RECEIPT OF STAMPED ENVELOPE.

Church Congress Report.

THE PRESENT TRIAL OF FAITH:
BEING
SERMONS
PREACHED IN ST. MARTIN'S CHURCH, LEICESTER,
BY
DAVID J. VAUGHAN, M.A.,
HONORARY CANON OF PETERBOROUGH CATHEDRAL; VICAR OF ST. MARTIN'S, LEICESTER; AND FORMERLY FELLOW OF TRINITY COLLEGE, CAMBRIDGE.

PRICE, NINE SHILLINGS.

MACMILLAN & CO.

LIFE IN CONSTANTINOPLE.

"THE TCHERKESS AND HIS VICTIM."
SKETCHES OF SOCIAL, MORAL, AND POLITICAL LIFE IN CONSTANTINOPLE.

BY A RESIDENT OF THE LAST THREE YEARS.

Crown 8vo. Cloth, 7s. 6d.

"Its variety is one of its charms. It will attract and give pleasure to many a reader who would have felt himself unequal to the effort of dealing with a disquisition on the state of affairs in Constantinople. The reader who honestly desires to have the opinion of an eye-witness on the actual condition of the Turkish Capital has not far to seek for his gratification. The author's views are, in the main, sound and sensible; he recognizes the good points as well as the bad points in the Turkish character. The Tcherkess and his Victim is an interesting book."—*Athenæum*.

"It will naturally claim attention at a moment when so much interest is centred on the action of the Porte. As its fidelity for the truth is vouched for it may be regarded as a fairly accurate portrayal of what is going on day by day in the Moslem capital."—*Publishers' Circular*.

CHURCH RAMBLES & SCRAMBLES,

BY A PERAMBULATING CURATE.

Crown 8vo; cloth extra; shortly.

JOHN HODGES, 24 KING WILLIAM STREET, W.C.

Advertisements.

IMPORTANT ECCLESIASTICAL PUBLICATIONS
(ALL COPYRIGHT),

Edited by Mr. W. EMERY STARK,

(Associate of the Institute of Actuaries, F.R.G.S., &c., &c.,

AND ISSUED BY

MESSRS. W. EMERY STARK & CO.

1.—For Patrons of Church Preferment.

"THE PRIVATE PATRON'S GAZETTE" (5th Edition—entirely new work—30 pages), containing full particulars (without, of course, names) of the requirements of about *160 bona fide Purchasers of Advowsons*, Next Presentations, Episcopal Chapels, together with remarks upon the present value of Church Property, sent only to *bona fide* Private Patrons, on receipt of Six Stamps. Circulation of last Edition 2000.

> *This work should be referred to by every Patron desirous of effecting a Sale, by private treaty, of any Church preferment.*

2.—For Purchasers of Church Property.

FIFTEENTH YEAR OF ISSUE.

"THE CHURCH PREFERMENT GAZETTE" (Monthly), containing full particulars, viz.: Diocese, Population, Income, Out-goings, size of House, prospect of Possession, and price of *Advowsons, Next Presentations, &c., for Sale*, by private treaty only, in almost every County and Diocese, together with remarks upon the present value of Church Property. Sent to Principals or their Solicitors only, on receipt of Six Stamps. Circulation about 1500.

> *This work will be found a most invaluable guide to all intending purchasers.*

3.—Exchanges.

FIFTEENTH YEAR OF ISSUE.

"THE BENEFICE EXCHANGE GAZETTE" (70 pages—largest and most valuable work of its kind ever issued), containing full particulars of Benefices for Exchange, town and country Parishes, in every County and Diocese, sent on receipt of Six Stamps. Every Incumbent *bona fide* desirous of exchanging should carefully peruse this work. Circulation last year—1550.

4.—For Incumbents.

FIFTEENTH YEAR OF ISSUE.

"THE INCUMBENT'S REGISTER," issued the 2nd week in each month, containing full particulars of about 200 Clergymen and candidates for Holy Orders (*all possessing good personal references*) *seeking sole Charges, Curacies*, and Temporary Duty. Sent on receipt of Six Stamps. Circulation about 4000.

> *This work has been for many years the recognised medium for Incumbents obtaining really satisfactory help, and references can be given to hundreds of Incumbents who have engaged Curates through the Office.*

5.—For Curates.

FIFTEENTH YEAR OF ISSUE.

"THE CLERICAL REGISTER," issued 1st of every month, containing particulars of VACANT *Sole Charges*, Curacies, and Titles with good stipends, town and country Parishes in every County and Diocese. Sent on receipt of Six Stamps. Circulation last year—about 5000.

> *This work should be carefully referred to by every Clergyman seeking a good position*

Address : Messrs. W. EMERY STARK & Co., St. Paul's Chambers, No. 23, Bedford Street, Strand, London, W.C.

REFERENCES *can with pleasure be given to Colonial Bishops, Canons, Archdeacons, and hundreds of Incumbents, who have for years entrusted their clerical business to* Messrs. W. EMERY

JOHN HODGES' NEW LIST.

LIFE IN CONSTANTINOPLE. THE TCHERKESS AND HIS VICTIM. Sketches of Social, Moral, and Political Life in Constantinople. By a Resident of the last Three Years. Crown 8vo., cloth, 7s. 6d.
[Now Ready.

A DEFENCE OF THE LITURGY OF THE CHURCH OF ENGLAND. By AMBROSE FISHER, "The Blind Scholar" of Westminster, sometime of Trinity College, Cambridge; ob. 1617. Edited, with Notes and Introductory Notice, by THOMAS BRYANT, Author of "Edmund, King and Martyr," &c.
[In the Press.

A CHRONICLE OF THE ENGLISH BENEDICTINE MONKS, from the Renewing of their Congregation in the days of Queen Mary to the death of James II.: being the Chronological Notes of DOM. BENNETT WELDON, O.S.B., a Monk of Paris. Edited, from a Manuscript in the Library of St. Gregory's Priory, Downside, by A MONK OF THE SAME CONGREGATION. Demy 4to., handsomely printed.
[In the Press.

*** A full prospectus post free on application.

THE EUCHARISTIC MANUALS OF JOHN AND CHARLES WESLEY. Reprinted from the Original Editions of 1748-57-9. Edited, with an Introduction, by the Rev. W. E. DUTTON, Vicar of Menstone. Sq. 16mo., cloth, red edges, 2s. 6d.; or in two parts, 1s. each.

THE HELIOTROPIUM; or Conformity of the Human Will to the Divine, Expounded in Five Books. By JEREMY DREXELIUS. Translated from the Original Latin by R. N. SHUTTE, B.A., with a Preface by the late Right Rev. A. P. FORBES, D.C.L., Bishop of Brechin. Square crown 8vo., second and cheaper edition.
[In the Press.

LIFE OF THE GOOD THIEF. From the French of the ABBE GAUME, Protonotary Apostolic. Fcap. 8vo.
[In the Press.

CORNELIUS A LAPIDE'S COMMENTARY ON THE GOSPELS. Translated by T. W. MOSSMAN, B.A., Oxford, Vol. 3, demy 8vo., completing S. Matthew and S. Mark's Gospels.
[Nearly ready.
Vol. 4, containing S. Luke's, and Vol. 5, S. John's Gospel.
[Preparing.

THE OFFICIAL REPORT OF CHURCH CONGRESS HELD AT OXFORD IN JULY, 1862. President, The Right Rev. SAMUEL WILBERFORCE, D.D., Lord Bishop of Oxford, Chancellor of the Most Noble Order of the Garter, Lord High Almoner to the Queen, &c. Edited, with a new Preface, containing a Sketch of the Rise, Progress, and Present Aspect of Church Congresses, by P. G. MEDD, M.A., Rector of North Cernay, and Examining Chaplain to the Bishop of St. Alban's, late Senior Fellow and Tutor of University College.

THE SAILORS' MANUAL: Being a Complete Manual of Instruction, Prayers, and Devotions for the use of Sailors and others engaged on the Sea. Compiled from various sources. Edited by the Rev. Canon SCARTH, M.A., Vicar of Holy Trinity, Gravesend, and Secretary to the Waterside Mission. Demy 16mo.
[In the Press.

THE CHURCHMAN'S BIRTHDAY TEXT BOOK. Royal 32mo. cloth; also in various bindings.
[In the Press.

THE PARSON'S SONS (a Novel). By ONE OF THEM. (ALTER BROWN). Crown 8vo., 1 vol.
[In the Press.

*** This is the first of a proposed Series of Original Novels by Eminent Authors. Each complete in one volume.

Advertisements.

MESSRS. BAGSTER'S LIST.

THE BURIALS ACT.
Free Church Services.
FOR MARRIAGES, BURIALS, BAPTISMS, AND THE LORD'S SUPPER.
Adapted from the Book of Common Prayer, for the use of Nonconforming Congregations.
Crown 8vo., French morocco, gilt edges, Price 2s.

PRESENTS FOR MINISTERS.

THE BLANK-PAGED BIBLE. The Holy Scriptures of the Old and New Testaments, with copious References to Parallel and Illustrative Passages, and the alternate pages ruled for MS. Notes. 8vo., in a strong Persian morocco binding, or in morocco, gilt edges, 21s.

The feature of this book is, that the ruled page is always on the right-hand side, thus facilitating the entry o M.S. notes.
"The practice of interleaving the Bible has never been more ingeniously carried out than in the plan developed by Messrs. Bagster. In a volume of not more than ordinary bulk, beautifully printed in clear and legible type, we find the advantages above detailed united in a manner highly valuable and practically useful. To the Biblical Student, and especially to the clergy, this Bible must prove of the highest value."—*Church of England Quarterly.*
"Among the many forms of beauty under which the Bagsters send forth Bibles to the world, we doubt whether there has yet been one that has more happily united usefulness and elegance."—*Christian Times.*
"Simple as this contrivance seems, it does great credit to the inventive ingenuity of the publishers in meeting the almost undefined wants of Biblical Students, to whom this beautiful volume will form a priceless acquisition."—*Journal of Sacred Literature.*

THE GREEK SEPTUAGINT, with an English Translation, and with Various Readings and Critical Notes. A NEW EDITION, specially prepared for Students. 4to., cloth, 12s.; morocco, gilt edges, 21s.; with the APOCRYPHA, in One Vol., 1,400 pages, cloth, 16s.

This is the only Edition of the Greek Septuagint version with an English translation side by side. An historical account of the version is given in the introduction, and this, with the various Readings and Notes, makes the work one of surpassing value to those who are studying the Greek Text.
"Both Greek and English are beautifully printed in type that is very clear, yet so small that the volume is quite convenient and handy in size."—*Guardian.*
"There is every reason why this version of the Hebrew Scriptures should be carefully studied, and to those who are thus minded the Messrs. Bagster's Greek and English Septuagint will be a most acceptable acquisition."—*Nonconformist.*

STUDIES ON THE TIMES OF ABRAHAM. By the Rev. HENRY GEORGE TOMKINS, Member of the Society of Biblical Archæology, &c. Profusely Illustrated in Chromo-lithography and Photo-tint. 4to., cloth extra, 16s.

This work is not a new biography of Abraham, but gives an account of the civilised world in which he lived, from Elam on the East to Egypt on the West, drawn from existing results of Egyptological and Assyriological research, and elucidates the true position and character of the Patriarch. The illustrations are chosen from an ethnographic point of view, as specimens of the leading races of the early world.
"Mr. Tomkins has produced a very vivid and truthful picture of the surroundings of Abraham."—*Academy.*

THE COMMENTARY WHOLLY BIBLICAL: An Exposition of the Old and New Testaments in the very Words of Scripture. Three Vols., 4to., cloth, marbled edges, 31s. 6d.

RECORDS OF THE PAST: Being English Translations of the Assyrian and Egyptian Monuments. Published under the sanction of the Society of Biblical Archæology. Edited by Dr. BIRCH. Cloth, 3s. 6d. Volumes I. to IX. now ready; Vol. XII. *in the press.* The Vols. may be had separately. *See List of Contents.*

BABYLONIAN LITERATURE. Lectures delivered at the Royal Institution. By Rev. A. H. SAYCE, M.A., Professor of Comparative Philology, Oxford; Author of "An Assyrian Grammar," "The Principles of Comparative Philology," &c. Demy 8vo., cloth, 4s.

DAILY LIGHT ON THE DAILY PATH: A Devotional Text Book for Every Day in the Year, Morning and Evening, in the very Words of Scripture. Two Editions, prices from 1s., may be had of all Booksellers.

ORIENTAL RECORDS. MONUMENTAL. Confirmatory of the Old Testament Scriptures. A Collection of the most important Recent Discoveries, especially in Western Asia and Egypt, derived from the highest attainable antiquity; confirmatory and illustrative of the statements of Holy Scripture. Illustrated. By WILLIAM HARRIS RULE, D.D. Crown 8vo., cloth extra, 5s.
Ditto. HISTORICAL. Crown 8vo., cloth extra, 5s.

SAMUEL BAGSTER & SONS, 15, Paternoster Row, London.

Church Congress Report.

JOHN HARDMAN & Co., BIRMINGHAM.

London Offices and Show Rooms—13, KING WILLIAM STREET, STRAND, W.C.

Chalices, Flagons, Cruets, Communion Services, Altar Candlesticks, Lecterns, Book Desks, Gas Standards, Coronæ, Candelabra, Candle Brackets, Candle Standards,

Altar Crosses, Flower Vases, Alms Dishes, Memorial Brasses, Carved Oak and Wrought Metal Chancel Screens, Communion Rails, in Brass, Wrought Iron, and Oak.

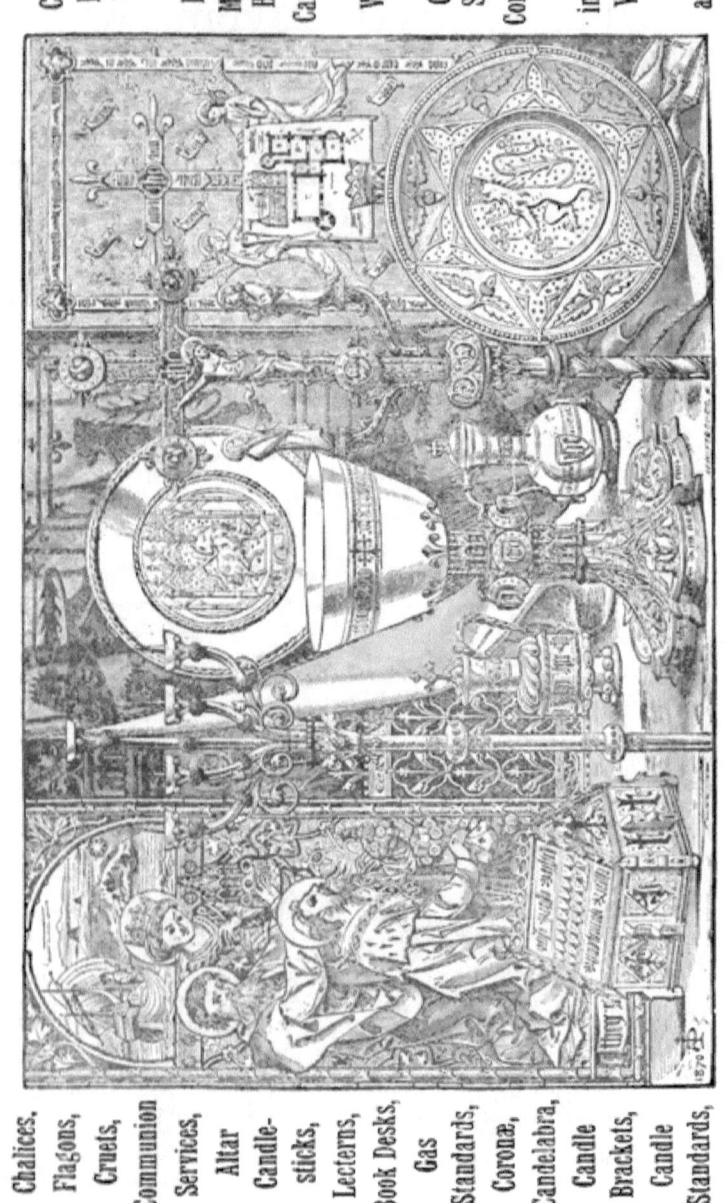

ARTISTS IN STAINED GLASS. Established 40 Years.

PRIZE MEDALS—London, 1851; Paris, 1855; Paris, 1862; Paris, 1867; Rome, 1870; Philadelphia, 1876; Paris, 1878-9.

An Illustrated Price List, Estimates, and Designs furnished on application.

www.ingramcontent.com/pod-product-compliance
Lightning Source LLC
Chambersburg PA
CBHW020058170426
43199CB00009B/325